ZEN
AND
THE ART OF
SAVING
THE
PLANET

THICH NHAT HANH

EDITED AND WITH COMMENTARY
BY SISTER TRUE DEDICATION

AFTERWORD BY SISTER CHAN KHONG

HarperOne
An Imprint of HarperCollinsPublishers

HarperCollins books may be purchased for educational, business, or
sales promotional use. For information, please email the Special Markets
Department at SPsales@harpercollins.com.

FIRST HARPERCOLLINS PAPERBACK PUBLISHED IN 2022

Designed by Bonni Leon-Berman
Calligraphy by Thich Nhat Hanh
© Plum Village Community of Engaged Buddhism, Inc.

Library of Congress Cataloging-in-Publication Data is available upon request.

ISBN 978-0-06-295481-7

22 23 24 25 26   LSC   10 9 8 7 6 5 4 3 2 1

CONTENTS

PREFACE

Sister True Dedication

Thich Nhat Hanh (or "Thay," as we call him) is a poet, scholar, peace activist, Zen master—and man of action. He embodies an inspiring, decisive, compassionate, and fearless engagement that springs from a place of serenity and insight. Thay teaches that to practice meditation is "to look deeply into the heart of reality, to see things that others can't see." And, as he says, "Once there is seeing, there must be acting. Otherwise, what's the use of seeing?"

A monk for nearly eighty years, Thay has found remarkable ways to combine his practice of meditation and mindfulness with extraordinary actions for peace and social justice, investing his life energy in training the next generation of engaged Buddhists, and building healthy communities of mindful living that can continue to be catalysts for change in the world.

In the 1960s, Thay created a movement of thousands of young

social workers in Vietnam before leaving for the West to call for peace. A leading voice for nonviolent social change, he collaborated with Dr. Martin Luther King Jr., with whom he shared a vision for building a "beloved community" that could transcend division, discrimination, and hatred—a community in which true reconciliation can be possible, among all people and among all nations. In the 1970s, together with friends and colleagues, Thay rescued boat people from the high seas off Singapore and initiated one of the very first international environmental conferences in Europe. Over the following decades, Thay created a way of teaching and applying mindfulness in everyday life that could be accessible to millions. He has shared his vision for compassionate leadership with politicians, businesspeople, teachers, activists, and more recently, Silicon Valley CEOs. And from his direct—and often painful—personal experience of unstable and polarized times, he has developed a simple yet powerful code of global ethics that offers a bright compass to guide our way forward.

In this very moment, we face a potent intersection of crises: ecological destruction, climate breakdown, rising inequality, exploitation, racial injustice, and the lasting impacts of a devastating pandemic. The situation is beyond urgent. In order to face these challenges with the best we've got, we need to find ways to strengthen our clarity, compassion, and courage. Cultivating a strong training in meditation and mindfulness is not an opiate to escape what's going on but a way for us to truly still the mind and look deeply, in order to see ourselves and the world clearly. From this foundation of clarity and insight we'll be able to take the most appropriate, effective action to transform the situation and create a regenerative culture in which all life is respected.

Thay tells us that "the world doesn't need another ideology or doctrine, but the kind of awakening that can restore our spiritual strength." This book, edited by his students, offers his most inspirational and timely teachings for the next generation and his guidance on how we can truly sustain our efforts to help our society and planet *without burning out*. Thay first asked us to begin working on this book almost a decade ago, and we are excited to have finally assembled in one volume his powerful Zen teachings on deep ecology, engaged action, community-building, and collective awakening, drawn from his writing, talks, interviews, and question-and-answer sessions. Thay gives voice to a very practical, everyday ethic that can guide our decisions and actions, transform the everyday habits that hold us back, and help us touch joy and meaning right in the heart of each moment. Thay explains that without such an ethic—without a spiritual dimension to guide our daily lives—we will lose everything.

Before Thay suffered a major stroke in 2014, many of us who lived and trained with him had a chance to experience his guidance directly. He nurtured and challenged us, encouraged and sometimes scolded us. He was tender as a grandfather, fierce as a warrior. There were times he mobilized us to be some of his many arms of engaged action in the world. And, whatever the task, it was always to be done with direct immediacy. (I learned that a young student should never ask her teacher, "Are you sure?") In true Zen style, there were moments when Thay declared, "Don't just *do

something, *sit there*!" At other moments, he'd call our names and send us off from our cushions, out of the meditation hall, to work on something urgent we'd not yet finished. There were days when the action was so pressing he'd remind us, with a gentle smile and a glint in his eye, that "there's no need to eat lunch. The human body can survive several days without food." And there were yet other days when, seeing us working so hard we'd forgotten to eat, he quietly went into the kitchen himself to prepare us hot soup for dinner.

It is hard to put into words Thay's compassion and light and his bright, penetrating gaze. It is hard to express his gentleness and warmth. It is hard to explain how much love and trust he offers unconditionally to all those who consider themselves his students. Thay encourages us to boldly reimagine an entirely new way of living and doing things, and to never be afraid to dream. And he reminds us, no matter what, to always work together, never alone. As companions on the path, we invite you to join us on a journey to the heart of his teachings on Zen and the Art of Saving the Planet.

I have been looking for you, my child,

Since the time when rivers and mountains still lay in obscurity.

I was looking for you when you were still in a deep sleep,

Although the conch had many times

Echoed in the ten directions.

From our ancient mountain I looked at distant lands

And recognized your steps on so many different paths.

Where are you going?

In former lifetimes you have often taken my hand

And we have enjoyed walking together.

We have sat for long hours at the foot of old pine trees.

We have stood side by side in silence

Listening to the sound of the wind softly calling us

And looking up at the white clouds floating by.

You have picked up and given to me the first red autumn leaf

And I have taken you through forests deep in snow.

But wherever we go, we always return to our

Ancient mountain to be near to the moon and stars,

To invite the great bell every morning to sound,

And help all beings to wake up.

FROM "AT THE EDGE OF THE FOREST"

by Thich Nhat Hanh

now
is the time
this is
it

INTRODUCTION

The beauty of the Earth is a bell of mindfulness. If you can't see it, you must ask yourself why. Maybe something is blocking the way. Or maybe you are so busy looking for something else you can't hear the call of the Earth.

Mother Earth is saying, "My child, I'm here for you; I'm offering all this for you." It's true: the rays of sunshine, the singing birds, clear streams, the cherry blossom in spring, and the beauty of the four seasons—it's all there for you. And, if you can't see or hear it, it's because your mind is too full.

The Earth is telling you that she is there and that she loves you. Every flower is a smile of the Earth. She's smiling to you, and you don't want to smile back. The fruit in your hand—it might be an orange or a kiwi—is a gift from the Earth. But, if you don't feel thankful, it's because you're not there for the Earth, for life.

An essential condition to hear the call of the Earth and respond to her is silence. If you don't have silence in yourself, you cannot hear her call: the call of life. Your heart is calling you, but you don't hear. You don't have time to listen to your heart.

Mindfulness helps us stop the distraction and come back to our breathing. Paying attention only to the in-breath and out-breath, we stop our thinking and, within just a few seconds, we awaken to the fact that we are alive, we are breathing in, we are here. We exist. We are not non-existent. "Ahhh," we realize. "I am here, alive." We stop thinking about the past, we stop worrying about the future, we focus all our attention on the fact that we are breathing. Thanks to our mindful breathing we set ourselves free. We are free to be here: free from thinking, anxiety, fear, and striving.

When we are free, we can answer the call of the Earth. "I'm here. I'm a child of yours." We recognize that we are part of the wonder. And we can say, "I'm free: free from everything that is preventing me from being fully alive. And you can count on me."

When you wake up and you see that the Earth is not just the environment, the Earth *is* us, you touch the nature of inter-being. And at that moment you can have *real* communication with the Earth. That is the highest form of prayer. In that kind of relationship, you will have the love, strength, and awakening you need to change your life.

The truth is that many of us have become alienated from the Earth. We forget that we are alive, here, on a beautiful planet and that our body is a wonder given to us by the Earth and the whole cosmos. If the Earth has been able to offer life it is because she, too, has non-Earth elements in her, including the sun and stars. Humankind is made of stars. The Earth is not only the Earth but the whole cosmos.

Only when you have this right view, this insight, will discrimination no longer be there, and there will be deep communion, deep communication between you and the Earth. All kinds of good things will come from it. You transcend the dualistic way of seeing things: the idea that the Earth is only the environment, and that you are in the center; and that you only want to do something for the Earth so *you* can survive.

When you breathe in and become aware of your body, and look deeply into your body, and realize that you *are* the Earth, that your consciousness is also the consciousness of the Earth that can become a liberated consciousness, free from all discrimination and wrong views, you are doing what Mother Earth is expecting you to do: to get enlightened, to become a buddha, so you can help all living beings, not only on Earth but even, ultimately, on other planets.

My generation has made many mistakes. We borrowed this planet from you, and we've done it great harm and destruction. Giving it to you now, we're ashamed. It's not as we would wish it to be. You are receiving a beautiful planet that is damaged and wounded. We are sorry. As someone belonging to the older generation, I hope the young generation can step up as soon as possible. This planet belongs to you, to future generations. Your destiny and the destiny of the planet are in your hands.

Our civilization is a civilization of borrowing. Whenever we want something we can't afford, like a house or a car, we count on our body and our labor in the future to pay back the debt. We borrow and borrow without knowing if we can ever pay back. In this way, we have borrowed from ourselves, from our health, and from the planet. But the planet can't take it anymore. And we have borrowed too much from you, our children and grandchildren. The planet and future generations are also us; we are not separate. The planet is us, and you are also us. The truth is there is not much of ourselves left.

It is very important to wake up and see that we don't *need* to borrow anymore. What is available in the here and now is *already* sufficient for us to be nourished, to be happy. And that is the miracle of mindfulness, concentration, and insight: realizing we can be happy with the conditions that are already available, that we don't need to strive to get more, exploiting the planet as we do so. We don't need to "borrow" anything. Only with this kind of awakening can we stop the destruction.

It's not something that can be done individually. We have to wake up together. And, if we wake up together, then we have a chance. Our way of living our life and planning our future has led us into this situation. And now we need to look deeply to find a way out, not only as individuals but as a collective, a species. You can no longer count on the elder generation alone. I have often said that one buddha is not enough; we need a collective awakening. All of us have to become buddhas in order for our planet to have a chance.

Be
still
and
see

PART 1

RADICAL INSIGHT
A NEW WAY OF SEEING

Are You Sitting Comfortably?

Sister True Dedication (—T.D.)

Thay is blazingly clear: there's one thing we have the power to change, which will make all the difference, and that is our mind. Our mind is the instrument with which we engage and interact with the world; it holds our despair and fears, our hopes and dreams. Our mind's way of seeing determines the decisions and actions we take or avoid, how we relate to those we love or oppose, and how we respond in a crisis. In Buddhism we often say that with our mind we create the world. Our perceptions are conditioned by language and culture and by society's tendency to put reality into boxes and categories that simply don't fit. These discriminating labels limit our clarity and our action to protect the planet and prevent us from living in harmony with each other and with the world.

We may want the world to wake up and act. But what kind of awakening would actually be helpful? What do we need to wake up *to*?

Buddhism speaks of two levels of truth: the level of labels and appearances, often called "conventional truth," and the deeper level of reality, known as the "ultimate truth." Thay teaches us that, if we're going to help our society and planet, we need to wake up to what's going on at both levels of truth.

In many talks in Plum Village, the international practice center and monastery that Thay founded in southwest France, Thay taught us one of the most ancient and powerful texts in Zen Bud-dhism, the Diamond Sutra. It is the world's first treatise on deep ecology and a treasure of humanity's shared wisdom heritage. The sutra originated in the northeast of the Indian subcontinent, sometime between the second and fifth centuries. There's even a ninth-century scroll of the Diamond Sutra, printed on paper from mulberry bark and hemp, found in the remote Dunhuang caves, where the old Silk Road entered western China. It is the world's oldest dated printed book. On a teaching tour to London a few years ago, Thay took a few dozen of us with him to see the scroll at the British Museum. Our times make it possible for wisdom to transcend geography and generations.

As you'll discover in the following pages, the Diamond Sutra proposes a deep contemplation to give us a breakthrough in our way of seeing the world. It offers a four-part meditation to cut through the stories we tell about what life is and isn't in order to help us get closer to the deeper level of reality as it truly is. It's known as the Vajracchedika Sutra—the "thunderbolt" or "dia-mond" that "cuts through illusion." Applying the teachings of the Diamond Sutra can give us a vast source of energy and clarity to take the right kind of action.

It's unbelievably hard to stop and step back. It may even be scary. The fact is, it's rare to get a chance to challenge the deeply held beliefs society imprints on us. For that reason, you may like to read the following pages slowly, taking time to see how these insights may apply directly to your own life. You may like to go for a walk to create space to contemplate these ideas, or to take some notes in a journal as you go along. Thay always says, as the Buddha did, "Whatever you do, don't just take my word for it. Put it into practice and see for yourself."

Ready for some truth-telling?

Spring Thunder

Many of us are barely awake. We're living in the world, but we can't really see it; it's as though we're sleepwalking. To wake up first of all is to wake up to the beauty of the Earth. You wake up to the fact that you have a body and that your body is made of the Earth and sun and stars. You wake up to the fact that the sky is beautiful and that our planet is a jewel of the cosmos. You have an opportunity to be a child of the Earth and to make steps on this extraordinary planet.

Second, to wake up means to wake up to the suffering in the world. You wake up to the fact that the Earth is in danger and living species are in danger. You want to find ways to bring relief, healing, and transformation. This requires a tremendous source of energy. If you have a strong desire in you, a mind of love, that is the kind of energy that will help you do these two things: wake up to the beauties of the planet to heal yourself and wake up to the suffering of the world and try to help. If you have that source of strength in you, if you have that mind of love, you are what can be called *a buddha in action*.

If you see the suffering in the world but you haven't changed your way of living yet, it means the awakening isn't strong enough. You haven't really woken up. In Zen, sometimes a teacher will shout, or hit you, so you can wake up—they'll do whatever it takes. The Zen master's shout is like a crash of spring thunder. It wakes you up and, with the rains that follow, grasses and flowers will bloom.

We need a *real* awakening, a real enlightenment. New laws and policies are not enough. We need to change our way of thinking and seeing things. This is possible; the truth is that we have not really tried to do it yet. Each one of us has to do it for ourselves. No one else can do it for you. If you are an activist and you're eager to do something, you should begin with yourself and your own mind.

It's my conviction that we cannot change the world if we're not able to change our way of thinking, our consciousness. Collective change in our way of thinking and seeing things is crucial. Without it, we cannot expect the world to change.

Collective awakening is made of individual awakening. You have to wake *yourself* up first, and then those around you have a chance. When we ourselves suffer less we can be more helpful and we can help others to change themselves too. Peace, awakening, and enlightenment always begin with you. You are the one you need to count on.

On the one hand, we must learn the art of happiness: how to be truly present for life, so we can get the nourishment and healing we need. On the other hand, we must learn the art of suffering: the way to suffer, so we suffer much less and can help others suffer less. It takes courage and love to come back to ourselves to take care of the suffering, fear, and despair inside.

To meditate is crucial, to get out of despair, to get the insight of non-fear, to keep your compassion alive so you can be a real instrument of the Earth helping all beings. To meditate doesn't mean to escape life, but to take time to look deeply.

You allow yourself time to sit, to walk—not doing anything, just looking deeply into the situation and into your own mind.

Eternity in the Present Moment

The extinction of species is taking place every day. Researchers estimate that every year over twenty thousand species go extinct, and the rate is accelerating. This is what is happening now; it's not something in the future. We know that 251 million years ago there was already global warming caused by gigantic volcanic eruptions, and that warming caused the worst mass extinction in our planet's history. The six-degree Celsius increase in global temperature was enough to wipe out 95 percent of the species that were alive. Now, a second massive warming is taking place. This time there is also man-made deforestation and industrial pollution. Perhaps within a hundred years there may be no more humans on the planet. After the last mass extinction, it took the Earth 100 million years to restore life. If our civilization disappears, it will take a similar time for another civilization to reappear.

When we contemplate this, it is only natural that a feeling of fear, despair, or sadness may arise. That is why we have to train ourselves to touch eternity with the practice of mindful breathing, with our in-breath and out-breath. Mass extinction has already happened five times, and the one underway now is the sixth. According to the deepest insights of Buddhism, there is no birth and death. After extinction, life will reappear in other forms.

You have to breathe very deeply in order to acknowledge the fact that we humans may one day disappear.

How can we accept that hard fact and not be overwhelmed by despair? Our despair is fueled by views we have about ourselves and the world. When we start to re-examine our views and change our way of thinking and seeing things, it becomes possible to transform the mind of discrimination that is at the very root of our suffering.

It is possible to train ourselves to see and experience the present moment in a deeper way. And once we touch reality deeply in the present moment, we touch the past, we touch the future, and we touch eternity. We are the environment, we *are* the Earth, and the Earth has the capacity to restore balance, even if many species must disappear before balance is restored.

It doesn't take years of practice to touch eternity in the present moment. In a split second, you can touch it. Taking just one breath, or one step on the Earth, with mindfulness and concentration can help you transcend time. When you touch the present moment deeply, you have an eternity to live.

Zen Roots

The Sanskrit term for meditation is *dhyāna*. The Chinese pronounce it *chan*, in Vietnamese we say *thiền*, and in Japanese they say *zen*. The Chinese character 禪 literally means "the practice of reflecting." In my tradition we use the expression "the practice of looking deeply."

In order to look deeply you need to make time to be there, with mindfulness and concentration, so you can direct your attention to what is going on and take a deep look. With the energy of mindfulness and concentration you can get a breakthrough and begin to see the true nature of what is there. It may be a cloud, a pebble, or another human being, or it may be your anger or even your body. And so, the practice of Zen, dhyāna, meditation, is to be fully present and look deeply.

Buddhism in Vietnam began with the meditation tradition. In the beginning of the third century there was a merchant from Sogdia in central Asia who traveled to what is today northern Vietnam, perhaps along what's called the Maritime Silk Road. He stayed there to do business and waited until the winds were favorable to sail back to India. This young merchant found Vietnam very pleasant, so he settled there and married a Vietnamese young lady. They had a little boy, half Indian and half Vietnamese, who would go on to become the first teacher of Buddhist meditation in Vietnam and China: Master Tang Hoi.

When Tang Hoi was ten years old, his father and mother passed away, and he was taken in by an Indian Buddhist temple, in what is now northern Vietnam, to train to be a monk. The temples had been established by Indian monks in the ports and centers of trade for the Indian merchants staying for extended periods. By the third century, Buddhism was flourishing, and as a young monk Tang Hoi studied both Sanskrit and Chinese. He established a community and taught in Vietnam before heading north across the border into the Wu Kingdom (now China) to share the practice of Buddhist meditation.

It is recorded that, when Tang Hoi came to the Wu Kingdom, there were not yet any Buddhist monks; he was the first. He set up a little hut and practiced walking meditation, and word of his presence began to spread. He was summoned by the king, who was very impressed and, around the middle of the third century, allowed Tang Hoi to build the first Buddhist temple in the Wu Kingdom. It became known as "The First-Built Temple," and, if you go to Nanking today, you can still see its ruins. There, Tang Hoi began to teach meditation and organized ceremonies to ordain the first Buddhist monks in China, about three hundred years before Bodhidharma.

Many people think of Bodhidharma as the first teacher of Zen Buddhism in China, but it's not true. Three centuries earlier, Tang Hoi was already teaching there. He is truly the first Zen master of Vietnam and China. And, while Bodhidharma did not leave behind any writings, Tang Hoi left behind many works that are still preserved, including precious translations and commentaries. He translated and taught the Diamond Sutra, one of the most beloved scriptures in the Zen tradition and the earliest text to explore deep ecology.

When we hear about the Diamond Sutra, we can imagine a Zen monk like Master Tang Hoi walking with a staff and carrying an old scroll in his bag.

Diamond Breakthrough

In the Diamond Sutra the meditator is urged to *throw away*, to release, four notions in order to understand our own true nature and the true nature of reality: the notion of "self," the notion of "human being," the notion of "living beings," and the notion of "life span." In the sutra it says that, if you are still caught in these notions, you are not yet free and you cannot be a real bodhisattva, an awakened being helping to relieve suffering in the world. But, if you can break through these ideas, you'll have the insight, understanding, and freedom you need to help save the planet.

It takes insight and courage to throw away an idea. If we have suffered deeply, it may be because we held on to an idea we weren't able to release. Throwing away is a very strong term. It's not just "letting go." All those centuries ago, it was Master Tang Hoi who used the term "throw away" to translate the Pali term *paṭinissagga*.

The purpose of looking deeply and meditating is to get insight, and insight is something we have to experience *for ourselves*. So, we shouldn't waste time accumulating new ideas and knowledge; we have to learn in such a way that helps us overcome our real challenges and obstacles. The aim of a Zen master is to help students to transform; it's not to transmit knowledge or views. A Zen master is not a professor.

My tradition belongs to the lineage of the ninth-century Zen master Linji. He said, "My aim is not to give you knowl-

edge. My aim is to help you break free from your views."
Understanding should not be only empty knowledge, but
deep insight. Insight is not the outcome of thinking. Insight is
a kind of direct intuitive vision that you get from strong concentration. It's not a product of thinking. It is a deep intuition.
And, if it is a real insight, it will have the power to free you
from your anger, your fear, your suffering.

Being able to see just once in a lifetime is no small accom-
plishment. If you've seen once, you can see again. The question is whether you have the determination and diligence.

You Are More than You Think

The first notion you must throw away is the notion of self. This
is a very deep-rooted belief in every human being: that there
is a self, separated from the rest of the world; that we're ourselves, and everyone and everything else, including the Earth,
is not "us." We're born with this strong belief that we're separate: "I am not you. That is your problem; it's not my problem." Intellectually, we may know that nothing can exist by
itself alone, but in reality, we still *believe* things can, and still
behave as if we're a separate self-entity. This is the base of our
thinking and behavior, and it creates a lot of suffering. It takes
intensive training to throw this notion away.

In fact, there *is* no one, no self, there. There is thinking; there is reflection. But there's not a person behind it. When Descartes said, "I think, therefore I am," he was saying that, during the time he's thinking, he is the thinking. The Buddha said that there is thinking going on but that it's not certain there's an "I" behind the thinking. The thinking is going on; that is something we acknowledge. But can we say there's a thinker? If there's a painful feeling, we can say for sure that there is a painful feeling going on. But, as for the person who is the feeler, that's not so sure. It's similar to saying "it's raining." The rain is something certain; it is raining, but there is no rainer. You don't need a rainer for the rain to be possible. And you don't need a thinker for the thinking to be possible. You don't need a feeler for feeling to be possible. That is the teaching of non-self.

In the idea of "self," there is the idea that I am this body, this body is me; or this body is mine, it belongs to me. But this notion does not correspond to reality. When you look deeply into your body, you see that your body is a stream. You can see your parents and ancestors there in that stream. So, the stream is there, but it's not sure that there is someone called "myself." And, in that stream, you can see ancestors and everything—not only human ancestors, but animal, plant, and mineral ancestors. There's a continuum. Whether there is a person, an actor, behind it is not so sure.

A better statement would be "I *inter*-am." It's closer to the truth in the light of interconnectedness, interbeing. If father and son, mother and daughter, have the insight of no-self, they can look at each other in the light of interbeing and there'd be no more problems. We inter-are. I *am* like this because you *are* like that.

It is very important to throw away the notion "I am" because it does not reflect the true nature of reality.

The notion of a separate self is like a tunnel that you keep going into. When you practice meditation, you can see that there is the breathing but no breather can be found anywhere; there is the sitting but no sitter can be found anywhere. When you see that, the tunnel will vanish, and there will be a lot of space, a lot of freedom.

Who Am I?

I am the continuation of my parents. I am the continuation of my ancestors. This is very clear. I do not have a separate self. Looking into myself, I can see my father, my mother in every cell of my body. I can see my ancestors in every cell of my body. I can see my country, my people, in every cell of my body. I can see that I am made of many elements that can be described as non-me elements. I am made of non-me elements and, when these elements come together, they produce me. So, I am that. I do not have a separate existence. I do not have a separate self.

This is right view. Seeing reality in this way you are no longer lonely because you *are* the cosmos. You have this body, but you also have a cosmic body. The whole cosmos can be found in you. You have a cosmic body right here and right now, and you can talk to the cosmos in you. You can talk to your father in you, your mother in you, your ancestors in you. You are made of non-you elements. You are the continuation of parents, ancestors, stars, moon, sun, rivers, mountains. Everything is in you. So, you can talk to them and you know that you are the world. You are the cosmos. And this can be seen with meditation. When you are concentrated, you begin to see.

Suppose a wave appearing on the ocean asks herself, "Who am I?" If the wave has some time to get in touch with herself, she will find out that she is the ocean. She is a wave, but at the same time, she is the ocean. And she is not only this wave but she is the other waves also. So, she sees the connectedness, the interbeing nature between her and the other waves, and she no longer discriminates between self and not-self. It's very important that the wave realizes she has a wave body, but she also has her ocean body. When the wave recognizes her ocean body, she loses all kinds of fear and discrimination.

This is the *goodness* of meditation. It can help you touch your roots and free you from discrimination and fear. If you believe that you have a self—separated from your ancestors and the cosmos—you are wrong. There is a *you*, but you are made of non-you elements.

If you live mindfully and with concentration, you will touch more and more deeply the truth that is in you. And one day you will find that you are resting on the ground of the cosmos. In Christianity they speak of "resting in God." When the wave is resting in the ocean, she's at peace. When you are resting in your cosmic body, you're at peace. And, if you practice walking meditation, every step can help you touch that cosmic body, that ocean body, that makes you immortal. You are no longer afraid of dying. But many of us are too busy, and have no time to breathe and to walk in order to get in touch with our cosmic body, our true nature of no-birth and no-death.

Meditation can be very satisfying. You are searching for yourself. You are searching for meaning. And meditation is to have the time to look deeply, to listen deeply. When you do that, you can get in touch with your true nature, and you release all fear and discrimination.

Deep Ecology

The second notion the Diamond Sutra tells us to throw away is the notion of "human being." We know that *Homo sapiens* is a very young species on Earth. We arrived very late, and yet we behave as if we're the boss here. We believe ourselves to be exceptional. We think we have a right over everything else and every other species, as though they have been created for us. With this view we have done a lot of damage to the Earth. We want safety, prosperity, and happiness only for humans, at the expense of everything else. And yet, looking deeply, we see that humans are made only of non-human elements, including plants, animals, and minerals. Not only historically but in this very moment, we continue to inter-be with all the non-human elements within us and around us. It's very clear: without minerals, plants, and animals, how can there be human beings? If you remove or return all these elements, a human being cannot exist anymore. And yet we seek to protect and defend ourselves by destroying our non-us elements, including other species.

In daily life, we need to use words to identify and define things, but it's not enough to live like that. In contemporary logic and mathematics, they still use the "principle of identity": *A can only be A. A cannot be B*. But the Buddha proposed that, when you look deeply, you see that A *is not only A*. A is made only of non-A elements. Humans are made only of non-human elements. Humans are made of *all* our ancestors. The mountains, the river, the rose, the planet are all made of non-mountain, non-river, non-rose, non-planet elements. When we can see that, we are free. "Human" and "mountain" are only labels, designations, without any real substance. They don't have a separate existence. This is the sword of the dialectics of the Diamond Sutra: A is not A, and that is why it can truly be A.

Man is present in all things, and all things are present in man. There is a mountain in us, do you see? There are clouds in us, do you see? It's not only that we *were* a cloud or a rock in the past, but we *are* still a cloud and a rock today. In former times we were also a fish, a bird, a reptile. We are a human being, yes, but at the same time we are everything. Seeing this, we know that to preserve other species is to preserve ourselves. This is interbeing, the deepest teaching of deep ecology.

In the Zen world they say, "Before practicing meditation, I saw that the mountains were mountains and the rivers were rivers. While practicing, I saw that mountains were no longer mountains and rivers were no longer rivers. And, after practicing, I saw that the mountains were *really* mountains and the rivers were *really* rivers." Seeing in this way, there is freedom.

I know ecologists who are not happy in their relationships. They work very hard to protect the environment, and it has become an escape from their partners. But, if someone is not happy within themselves, how can they help the environment? That is why to protect non-human elements is to protect humans, and to protect humans is to protect non-human elements. The insight of interbeing has the power to wake us up.

Life Has No Limits

The third notion we have to break through is the notion of "living beings." Many of us are caught in a distinction between sentient or "animate" beings and non-sentient or "inanimate" matter. Yet the science of evolution tells us we have not only human and animal ancestors but also mineral ancestors. To separate out living beings from the inanimate world and make a divide between them is incorrect.

We are made of non-sentient elements. The speck of dust, the elementary particle, the quark—they are us and we are them. We need to transcend the notions of body and mind, matter and spirit, consciousness and the material world. They are a big obstacle. Scientists of our time have discovered that even photons and electrons have their own intelligence; it is no less than consciousness. They are not inert and lifeless. A kernel of corn has its own way of knowing: you just plant it in the soil and within ten days it knows how to sprout and go on to become a towering maize plant with leaves, flowers, and cobs. The so-called inanimate things are not so inanimate; they are very alive.

You can also translate the term "living beings" here as "mortals." We discriminate not only between "living" and "inanimate" but also between what is living—or "mortal"—and what is holy or immortal. We have a tendency to discriminate between living beings and holy beings. The meditation here is to look into yourself deeply and realize that you are made of non-you elements, *including the element of holiness*. We must remove the notion of a "living being" as different from a non-living being, or an enlightened, holy being, because that notion creates so much division, discrimination, and suffering. This is the teaching—the revolution—of the Diamond Sutra.

When we look at the Earth in the light of this insight, we don't see the planet as inert matter but as a sacred reality that we are also a part of. Seeing in this way, our attitude toward the planet will change. We'll be able to walk with steps of love and respect on the Earth, and we'll realize our great capacity to help.

Watch Out! Don't Get Caught

In Zen, there is a koan, a question for contemplation: "Does the dog have buddha-nature?" Not only does the dog have buddha-nature, so does the stone—and the planet. The Earth manifests insight, awakening, happiness, and many other virtues. The Earth is a female buddha, a mother. "Whose mother?" you may ask. The mother of human-form buddhas and non-human-form buddhas. When we are not caught in signs it is easy to recognize the presence of a buddha.

Whenever we use the word "buddha" it is merely a concept of a buddha. Perhaps you have already heard the Zen story about the word "buddha." A Zen master, while teaching, used the word "buddha," and as he did so he used it very carefully because the word "buddha" and the idea of "buddha" are tricky. They can become a prison for those listening. People may think they know who or what the buddha is, and they can get imprisoned in that idea.

Using the term "buddha" is very dangerous, just as using

the term "God" is also dangerous. So, in order to help his listeners not get caught, the Zen master said, "My friends, I'm forced to use the word 'buddha.' I don't like the word; I'm allergic to it. And every time I use it, I have to go to the river and rinse out my mouth three times." A very strong teaching. Very Zen. Everyone in the audience was silent. But there was one student sitting at the back, and he stood up and said, "Teacher! Every time I hear you say the word I have to go to the river and wash my ears three times!"

We are lucky that there was such a teacher and such a student to help us not get lost or caught in ideas and words.

Mindfulness, concentration, and insight are in us all as potentialities—seeds—and that is our buddha-nature. Everyone has buddha-nature; that is the good news. It's not hope; it is a reality. The root *budh-* in Sanskrit means "to wake up." If you wake up to the beauties of the planet, you are already a buddha. And, if you know how to keep that spirit of awakening alive all day, you're a full-time buddha. It's not so difficult to be a buddha.

You Are Non-Local

You may believe that you are only this body. But you are much more than this body. To meditate is to see that you are also over there and there and there—everywhere. Your nature is non-local.

27

Sometimes Zen masters have to make up new words, especially when old words become sick and lose their meaning. One of the terms coined by the ninth-century Zen master Linji is the phrase "true person" (眞人 in Chinese; *chân nhân* in Vietnamese). These two words are very important. He said we need to live and practice mindfulness in such a way that we reveal our true person, which cannot be found in space or time. It has no coordinates. Like an electron in physics, it cannot be grasped. Sitting here in this moment, we realize the planet is there, with her mountains, rivers, and sky, and we inter-are with all these elements. The clouds are there in the sky, but they are also in us. The light of the sun is shining out there, but it is also shining in us. Your true person is a person of wonders.

When catastrophes or natural disasters happen and thousands of people die, we may ask ourselves, "How can such a thing happen? Why do some people have to die, and not others? Why do I get to survive?"

I have practiced sitting and looking deeply, and what I saw is that, when they die, we also die with them. Because we inter-are with them. It is like when someone we love dies, a little part of us also dies; somehow, we die with them. Those who die in natural disasters have died for us, and we are alive for them. Depending on our way of living now, their death can mean something. As we continue to live, they continue to live with us. We carry them all in us. With this insight of interbeing, we can have peace.

You Are Non-Temporal

The fourth notion the Diamond Sutra teaches us to remove is the idea of a "life span." We believe that there is a moment in time in which we were born, and there's another moment in which we die, and we think we only exist between these two points. This is to be caught in the idea of a life span. We think we are only on this Earth for a very short moment. We have the impression that we came from the realm of non-being into being, and, after staying in the realm of being for perhaps a hundred years, we'll pass back into the realm of non-being again.

It is recorded in the Sutra of Forty-two Chapters that one day the Buddha asked his monks, "How long does a lifetime last?" And one of them said, "Oh, a hundred years." The Buddha smiled. Another monk said, "Twenty years." And another said, "One day." And then one monk said, "The length of a single breath." And the Buddha said, "Yes, that is correct." We are being reborn in every breath. Today's you is already a new lifetime succeeding yesterday's you.

There is a Zen koan that asks, "Where were you before your grandmother was born?" This is not a philosophical question but an object of contemplation that should capture your attention day and night. In order for it to be a real koan, you have to invest 100 percent of your flesh, bone, and mind to get the breakthrough.

Observation tells us it is impossible to pass from something into nothing, to go from someone to no one. There is only continued manifestation in different forms. Before we were born, we were already there, and, after we die, we will continue to be there. Nothing can pass from the realm of being into the realm of non-being. As the scientist Lavoisier discovered, "Nothing is created, nothing is lost, everything transforms."

For example, my father has passed away, but he has not disappeared. He is still available. In every cell of our bodies we have our parents and ancestors, and we can talk with them, here and now. I do that all the time. I feel that my father is always alive with me. I invite him to walk with me and breathe with me. And what he could not do in his lifetime I try to do for him. We inter-are.

Your teacher, who has transmitted things to you, is also in you, and you are transporting him to the future. Your teacher in you may look different, sound different, and feel different from your teacher outside. I have been engaging in conversation with my own teacher. I know that everything I do is for him too. I carry him into the future, and I transmit him to you so you can continue to carry him into the future. My teacher in me does not look exactly like my teacher outside. My teacher in me is more aware of what is going on in society and can devise new teachings to help the world. My teacher has done his best and I, as a teacher, have done my best, but there are things we have not done, and our students must do them for us. It's very kind of you to help your teacher in you to evolve.

In the same way, the buddha in us should also evolve. We can help him be more relevant. The buddha of our time knows how to use a cell phone but he is a free person when he uses the cell phone. The buddha of our time knows how to help address the challenges of our times so we won't destroy the beauty of our planet or waste time competing with each other. The buddha of our time would like to offer the world a global ethic so everyone can have a path to follow, a path that can restore harmony, protect the planet, prevent deforestation, and reduce emissions. You, as the continuation of the buddha, must help offer the world a path that can prevent the destruction of our ecosystems and reduce fear, violence, and despair. We have to allow our ancestors, our teachers, the buddha in us, to act. The work of saving the planet continues beyond our lifetime.

The Task of a Meditator

Throwing away these four wrong notions is the meditation proposed by the Diamond Sutra. It is a *samādhi*, a powerful kind of concentration that you must sustain in your daily life. While eating, walking, cooking, and sitting—whatever you are doing—you should train yourself to keep this insight of interbeing alive so you can break through the ideas of a separate self, a human being, a living being, and a life span. Doing so, you free yourself from discrimination—the discrimination that is the ground of *all* suffering.

In our mindfulness practice centers and monasteries, whenever we hear the sound of a bell, we return to our breathing, and we return to the present moment. We stop talking, we stop doing whatever we're doing, we relax our body, release any tension, and begin to follow our breathing. The bell— whether it is the great temple bell, the clock in the dining hall, or a bell of mindfulness on our phone or computer—gives us an opportunity to stop and look deeply. The simple act of breathing and listening to the bell is a way to train ourselves in the art of stopping. In just two or three breaths we wake up to what is going on inside us and around us. The bell invites us to see that we are the world, we are the cosmos; there is no separation. We embrace limitless space and infinite time, and that moment becomes an eternal moment; we do not lack anything. The past, the present, and the future are all contained in this moment.

All fear, anger, despair, and anxiety arise from our wrong notions. And, when we remove them, and see reality more clearly, we release our suffering in a very concrete way. If you can touch reality like that, you will have right view. And when you have right view, all your thoughts will be right, all your words will be right, and all your actions will be right. The deep meditation of the Diamond Sutra brings forth non-fear, non-anger, and non-despair—the strengths we need for our work. With non-fear, even when we see that a problem is huge, we will not burn out. We will know how to make small, steady steps. If those of us who work to protect the environment contemplate these four wrong notions—of a separate self, of human beings, of living beings, and of a life span—we will know how to be and how to act.

The insight of interbeing will help you feel much more at ease in yourself. Removing wrong views, throwing them away, and breaking through into the heart of reality are the tasks of a meditator.

Seeing and Acting in the Light of the Diamond Sutra

—T.D.

The Diamond Sutra is a kind of prism to help us reimagine who we are; it's a radical way to broaden our horizons about what life is and isn't—and about the true nature of the planet we'd like to protect. It gives us a chance to step back from politics, policies, and news and check our perspective on reality and the very foundation of how we perceive the world.

Meditating on these insights doesn't take place only in the sitting position. Preparing breakfast, taking a shower, walking down the street, and enjoying the sunset or night sky are all moments when we can hold the insight of interbeing in our heart, allowing ourselves to be fully present and our mind to be still. A flash of insight doesn't need hundreds of hours sitting on a cushion.

The Diamond Sutra helps us touch the deep realization that we are intimately bound to the web of life. With the insight of interbeing, we realize that we're never alone, we're never powerless, and everything we will ever do *counts*. This is a real source of comfort. The Diamond Sutra invites us to throw away the idea that we're separate from our family, friends, and colleagues, and to throw away the idea that we're separate from the planet. These teachings invite us to experience a very tangible interbeing, right in this very moment, and to touch a sense of identity that is far vaster than we usually allow. We get to break free not only from a feeling of separateness "in real life," but also from the pressures of our virtual online "self" or persona, allowing us to experience, in this very moment, "our true person," across space and time.

In its widest sense, the Diamond Sutra helps us transcend any tendencies to have a "superiority complex" about humans being exceptional or distinct from the world around us, as well as the tendency to have an "inferiority complex" about being too flawed by our human nature to be able to help. We are neither. Realizing this can be both humbling and empowering.

The Diamond Sutra challenges us to throw away the idea that our contribution begins and ends with this life span. When we can touch interbeing across the axes of space and time, our ancestors, our descendants, and all those separated from us by circumstance become *accessible* to us, and we can open up to receive their energy, solace, and support. By radically re-examining our ideas about a self, a human being, a living being, and a life span, we can begin to transform some of the feelings of despair that may be paralyzing us, and release an energy of vitality and non-fear. Can we hear the voices of previous generations, and of the next? Can we hear those voices in our own times that are not being heard? Can we hear the voices of other species, and of the Earth?

In Plum Village, after teaching the Diamond Sutra, Thay would lead us on a walking meditation outside. We'd walk out along muddy paths through the oak forests to admire the view across the rolling French countryside. Sometimes an old church bell would echo noon across the valley, and we'd all take a few minutes to stop and breathe with the sky, with the land, with our ancestors, and with whatever we're carrying in our heart. The present moment and eternity happen at the same time.

Whose eyes are enjoying the sunset? Whose feet are walking? How many generations of ancestors are walking with you? What's a lifetime? When did the chain of heat that is your bodily warmth first begin?

I remember one day twelve years ago, soon after I had ordained, we monastics were enjoying the fresh magnolia blossoms at Thay's hermitage. I sat down to rest on the grass, when suddenly Thay appeared beside me. I joined my palms in greeting and bowed. Thay tilted his head quizzically and inquired, "Who are you?" I was shocked. I knew there were a lot of us, but, since he'd given me my monastic name, I thought he might remember it. I smiled faintly, dumbstruck. In return, Thay smiled a forgiving smile, and then moved on. I realized I hadn't quite understood. Some time later, he stopped another sister in the meditation hall and asked the same question: "Who are you?" "I don't know!" she quipped back, and Thay broke out in a grin. "That's real communication between teacher and student!" he declared, delighted. Thay teaches us that the mind of not-knowing is a mind that's open, free, and awake to infinite possibility.

We may feel immense pressure to save the planet in this lifetime, and we may be afraid that we can never do enough. The stark truth is that the planet doesn't need to be saved only once; it needs to be saved countless times, for eons to come. It's impossible to save the planet once and for all, or on our own. That the planet can be here now is a miracle, born of countless favorable causes and conditions over billions of years. And the planet will continue to need countless favorable causes and conditions going forward. This realization is good news. We belong to a stream of life, and this moment is our time and our turn to do our part, and to do whatever we can to pass on what we learn to future generations, so they can do theirs. The central part of this book—*The Action Dimension: A New Way of Living*—will explore Thay's bold vision for how we can all participate in creating a truly regenerative and compassionate culture that can continue even beyond our own lifetime.

Some of us may be haunted in our dreams or on our screens by fears of an apocalyptic "end" to life on this planet. Grief, anxiety, and sorrow may imprint themselves in our chest, our brows, our minds, clouding our days and disturbing our nights. This is the suffering of our time. We may know people who ask, "How can *any-thing* I do in my lifetime *ever* be enough?" We may know others who say, "Nothing matters, everyone will die anyway, the Earth will crash into the sun in a few billion years, so why not just enjoy whatever we want?" The contemplations of the Diamond Sutra reveal that these perspectives are still caught in the notions of a separate self or life span. The insight of interbeing breaks through the idea that whatever happens to the Earth after we die has nothing to do with us. We can no longer say that we only want to do something for the Earth if we think our "self" will have a future in it. We and the Earth inter-are.

There are those of us who shrug cynically, saying, "What's the meaning of life, anyway?" As soon as we say something like that, we make the assumption that we know what "life" is, and the only problem is understanding its meaning. But are we so sure we know what life is? The insights of the Diamond Sutra show us that life is much more than we think. It is possible to train ourselves to see that in every moment we participate in the future of the Earth, and that everything we do today can contribute to the planet's health and vitality.

You Cannot Become Nothing

The Buddha said that one minute of contemplating death is very rewarding because, if you know what death is, you become more alive. When I was young, I thought to myself, "I'm young and full of life, why do I have to think about death?" But, later on, I discovered that, if you meditate on death, you cherish life more, and you can touch the joy of being alive.

A permanent, immortal soul is something that cannot be accepted, either by good Buddhists or good scientists; everything is impermanent and continues to manifest in different forms. But the opposite view, that after this body disintegrates you will disappear altogether, is another extreme, another wrong view: the view of annihilation. We shouldn't fall into the trap of eternalism, but we shouldn't fall into the trap of annihilation either and say that after death we become *nothing*.

Imagine for a moment that you are a cloud. This is a meditation. You are made of tiny crystals of ice or water. You're so light you don't fall, you float in the sky. There is interaction, collision, between all the trillions of tiny crystals. At one point, they may combine to become hail or rain and begin to fall. But halfway down they encounter hot air and evaporate again. And so, you are going up, falling down, and coming up again. Transmigration, reincarnation, and rebirth are continuously taking place in your cloud. As a cloud you don't need to become rain to have a new life; you have a new life every moment. You may think the cloud just floats there and is the same cloud, but it's not true. A cloud is very active and full of energy.

For us, as human beings, it is the same. Rebirth and continuation are always happening; every moment we are producing thoughts, speech, and actions. Our actions are an energy that has an effect on ourselves and on the world; they are our product. They are our rain, our snow, our thunder and lightning. In Buddhism the word for action is *karma*. It is a very important term.

What's Your Karmic Footprint?

Imagine part of your cloud becoming rain and falling down to join the river below. Being the rest of the cloud in the sky, we can see our continuation down there. Floating up here is nice, but flowing down there is also nice. So, we are both up here and down there. We're speaking of a cloud, but we can speak of human beings in the same way. We can see ourselves in the fruits of our actions, in our loved ones, in everything we have realized. That is the way to look: you see yourself not just in this body but everywhere, in your thoughts, speech, and action that continue you in the world. This is something wonderful about meditation. You realize you can do something to have a better continuation. It is possible to produce thoughts, speech, and action of compassion, understanding, and forgiveness. There is cause for hope, for joy.

There is a so-called "scientific view" that everything happens by coincidence. The British philosopher Bertrand Russell described man as "the product of causes which had no prevision of the end they were achieving . . . his origin, his growth, his hopes and fears, his loves and beliefs, are but the outcome of accidental collocations of atoms." According to this view, there is no intelligent design, just coincidence.

In Buddhism, we don't speak of a project or plan made by a god. But we do speak of a dynamic force that underlies everything and determines the state of the world, the state of the Earth, and that force is karma, action. The fate of the planet depends on our action. It doesn't depend on a god. It doesn't depend on chance. It depends on our true action. I believe that scientists will have no difficulty accepting this.

In Buddhism we always understand action as having three aspects: thinking, speaking, and acting. When we produce a thought, it is energy, it is action, and it can change us and change the world, in a good way or in a bad way. If it is the right kind of thinking, it will have a healing, nourishing effect on our body and on our environment. Right thinking can make the world a better place to live, just as wrong thinking can transform the world into a hell. Speaking can have the effect of releasing tension, reconciling conflict, or bringing hope. Or speaking can destroy hope or cause a family to break.

Speaking is a kind of energy, a force. And physical action is also an energy that can bring healing to ourselves and to the world; we can *do something* that has the power of protecting, of saving, of supporting and bringing relief.

It may be that yesterday we produced a thought of hate or anger, or we may have done something in hate or anger. And today we realize that it's not a very good continuation to have. It is very possible to do something to transform it. You establish yourself in the present moment, aware of your body and breathing. And you bring to mind that yesterday you thought, said, or did such a thing, which may have brought harm to yourself and others. Then, sitting here, firmly established in the here and the now, you produce another kind of thought in the opposite direction: a thought of forgiveness, compassion, understanding. As soon as that thought is produced, it will catch up with the other thought and neutralize it right away. That's changing the karma. It is possible, with the practice of mindfulness in the here and the now.

There is free will, and the possibility of transforming; there is probability. Free will is mindfulness; when mindfulness intervenes, we know what we are thinking, saying, and doing. If we like it, we allow it to continue. If we don't like it, we are free to do differently.

Every thought we produce, every word we pronounce, everything we do *changes* our body, mind, and environment. This impact is called "retribution." The environment in which we find ourselves is *us*, and it is the result of our action. We have been living in such a way that we have destroyed our environment and many species have disappeared. That *is* our retribution.

You have to be responsible both for your own body and mind and for your environment. Your environment is you. When you look at the tree, don't think that the tree is something other than you. The tree *is* you. What you produce in terms of thought, speech, and action are energies that can never be destroyed. With mindfulness, compassion, and understanding you can assure a better retribution in the future, for yourself and for the world.

Indestructible

I remember the intense days directly following the September 11th attacks in America. I was in California at the time, and the suffering, fear, and anger in the country were immense. Life seemed to have stopped. I was scheduled to fly from California to New York to give talks and lead retreats on the East Coast. On the plane, there was a lot of fear and distrust on everyone's faces. The pilot tried to make a joke to relax the atmosphere, but no one laughed.

I was due to give a talk at the Riverside Church on September 25th. The night before, brothers and sisters in my community came to sit with me, including many young monks and nuns. I shared with them that the atmosphere is full of hate, fear, and anger, and, although America is ready to do something to punish the perpetrators, I would be advocating non-action, calming, stopping, and non-violence. There was a concern that someone, out of anger, may attack or shoot me at the talk. It was a dangerous situation, and they were worried for my safety.

I told them that, even if I'm killed for teaching the Dharma—the truth—you still have me as a teacher. But, if I lack courage, if I refuse to share my insight and compassion, then my body may still be alive, but you wouldn't have a good teacher.

We went ahead with the talk. So many people came that there wasn't enough space for everyone in the church. After two and a half hours of being together, listening deeply and breathing mindfully, you could see the deep transformation on people's faces. They looked much better than when they came. They had got some relief and suffered much less. You need to have the gift of non-fear within you in order to be able to offer it to others.

Two Kinds of Truth

In Buddhism, we distinguish between two kinds of truth: the conventional (relative, historical) truth and the ultimate (absolute) truth. On the level of conventional truth, we differentiate mind from matter, son from father, humans from other species, being alive from being dead. But, on the level of the ultimate truth, such differentiation is not possible. The ultimate truth transcends the ideas of separate selves, separate species, and even the idea of "birth" and "death." At the level of the ultimate truth, there is no such thing as death; there is only continuation.

The ultimate truth is not something that is separate from the conventional truth. When we touch the conventional truth deeply, we touch the ultimate. If we touch a cloud superficially, we only see the existence and non-existence of the cloud; that this cloud is *not* other clouds. All these notions can be applied to a cloud. Using our mind of discrimination to look at a cloud in this way, we touch only the phenomenal aspect of the cloud. We're in the realm of conventional truth. But, using mindfulness and concentration to look at the cloud more deeply, we discover that the same cloud is free from birth and death, being and non-being, and we touch the ultimate truth of the cloud. We don't have to throw away the cloud in order to touch its true nature.

There is no conflict between the two kinds of truth; they are both useful. The knowledge of conventional truth can be practically applied in technology and in daily life. Every one of us needs a birth certificate to get an identity card or passport. Without that, we can't go anywhere. We can't say, "Well, my nature is the nature of no-birth and no-death. I don't need a birth certificate!" And when someone dies, we have to report that they have died. We cannot say, "He can never die, I don't need to declare his death."

In Buddhism, when investigating reality, we follow a principle called "the separate investigation of phenomena and noumena." It is similar to the different approaches in classical science and modern science: the methods you apply in classical science must be left behind as you enter the realm of quantum physics. When you want to approach the ultimate truth, and break free from all discrimination and wrong views, you have to leave behind all the words, notions, and ideas that helped you investigate the phenomenal world. This is very important.

If it is a conventional truth, mindfulness helps us know that it is a conventional truth, and we are not caught. And if it is an ultimate truth, mindfulness helps us know that it is the ultimate truth, and we are not caught in the ultimate truth either. So we are free. It is possible to accept both kinds of truth. We cannot claim that one truth is better than the other, or that one truth is the *only* truth.

Face Your Fears

The Buddha advises us to look deeply and directly into the nature of our fear and get acquainted with it. Most of us are afraid of dying, of getting sick, of getting old, and of being abandoned. We're afraid of losing what we cherish and losing those we love. Many people suffer deeply but don't even know they suffer. We try to cover up our suffering by being busy, not because we want to be busy but to avoid touching the suffering. We need to do everything we can to realize that suffering is there and to learn ways to take care of it.

We shouldn't try to run away from fear but take time to recognize it, embrace it, and look deeply into its roots. In Buddhism, we have a meditation to face and transform our fears, called "the Five Remembrances." Breathing mindfully, in and out, deep and slow, can give us stability as we contemplate:

1. *I am of the nature to grow old. There is no way to escape growing old.*
2. *I am of the nature to have ill health. There is no way to escape having ill health.*
3. *I am of the nature to die. There is no way to escape death.*
4. *All that is dear to me and everyone I love is of the nature to change. There is no way to escape being separated from them.*
5. *I inherit the results of my actions of body, speech, and mind. My actions are my continuation.*

The life of a civilization is like the life of a human being: at the level of appearances, it also has its life span and will have to end one day. Already there have been many civilizations that have been destroyed, and ours is no different. We know that if we continue to live the way we do, destroying our forests, polluting our waters and skies, disaster cannot be avoided. There will be catastrophes, floods, and new diseases and many millions of people will die.

If we continue to live the way we are living, the end of our civilization will be certain. Only by waking up to this truth as a human species will we have the *insight* and *energy* we need to change our way of life.

We have to learn to accept that the extinction of many species on Earth, including humankind, is possible. But, if humans have appeared on Earth once, humans can appear again. We can learn from Mother Earth's patience, non-discrimination, and unconditional love. We see that the Earth can renew herself, can transform herself, can heal herself—and can heal us. That is a fact. We have to think of time geologically. A hundred years is nothing. In the present moment, if we go deep, we can embrace the whole of eternity.

The Way Out Is In

When you can face the truth and fully accept reality as it is, you will have a breakthrough and be able to have peace. The truth is so obvious. But, if you continue to resist the truth and allow fear, anger, and despair to overwhelm you, you cannot have peace, and you won't have the freedom and clarity you need to help. If all of us panic, we will only accelerate the death of our civilization.

The way out is in. You have to go back to yourself, face your deepest fears, and accept the impermanence of our civilization. The practice is to deal with our fear and grief *right now*; our insight and awakening will give rise to compassion and peace. Otherwise, we'll only make ourselves sick with denial and despair. If you can be at peace in the face of the reality, then we have a chance.

When we look at the present moment, we can already see the future. But things *are* impermanent. Humanity *can* change. First of all, we have to change ourselves. And, if we *ourselves* can do our best, that gives us a lot of peace already. The future of the planet doesn't depend solely on one person, but you have done your part. And that is why you can have peace.

The fact is, we already have enough technological solutions. But we're so possessed by fear, anger, division, and violence that we are unable to make good use of them. We're not making the challenges a priority; we're not investing our time and resources; we're not collaborating. The big powers are still investing a lot of money in manufacturing weapons instead of investing in new sources of energy. And why do countries need weapons? Because they have fear; both sides have fear. So, we need to transform our individual and collective fear.

The problem is a human one. And that is why we need a

spiritual dimension. If you can generate the energy of calm, acceptance, loving kindness, and non-fear, you can help offer and introduce that dimension of non-fear and togetherness to the situation. Technology alone is not enough to solve the problem. It needs to go together with understanding, compassion, and togetherness.

Our spiritual life, our energy of mindfulness, concentration, and insight, is what can bring about the energy of peace, calm, inclusiveness, and compassion. Without that, I don't think our planet will have a chance. So, please, in your sitting meditation, in your walking meditation, in your contemplations, look deeply to get the insight of peace, acceptance, and non-fear. It needs to be a real insight. Our peace, strength, and awakening will bring us together, and each of us can do our part to avoid catastrophe and contribute to saving the planet.

Zen and the Art of Catching a Snake

—T.D.

In Buddhism, it's said that putting the teachings into practice requires the same kind of skillfulness as catching a snake: try to grasp it in the wrong way, and it'll turn around to bite you. The same is true for the deep teachings we've been exploring so far, including the teachings on the two levels of truth and the teachings on contemplating our own impermanence and the impermanence of our civilization.

And so, if contemplating the end of our own life or our civilization gives rise to a paralyzing feeling of despair or numbness, we're applying that contemplation in the wrong way. The visualization may be shocking; there may be resistance; there may be tears, anger, and frustration as we hold the possibility in our heart. But, ultimately, the intention is to break through to a new horizon of realism, possibility, and, as Thay says, *peace*.

It is very powerful to recite the Five Remembrances—on page 46—every night before we go to sleep. We can recall the words silently, taking a few steady breaths, and train ourselves to slowly digest every line. Each one is a real stone to swallow: I *will* get sick, I *will* get old, I *will* die, I *will* eventually be separated from those I love. And so, have I lived today in such a way that I can truly cherish everything I have and everyone I love? And how would I like to live tomorrow? What *matters* to me the most?

We've been exploring the apparent contradictions between the two levels of truth. The Five Remembrances say that there is such a thing as death, whereas the four teachings of the Diamond Sutra seem to suggest there isn't: life is without limits across space and time. In deep Buddhism, we learn that both truths are happening at the same time and that by deeply contemplating the conventional truth we can touch the ultimate. One leads to the other. It is by contemplating interbeing and the energy of our actions (karma) in the world that we can expand our vision and touch the ultimate truth beyond signs and appearances. By doing so, we can get the realization that *every* action of body, speech, and mind *does* have an impact; whatever we do resonates, far wider than we usually imagine. And that gives us a huge motivation to take care of the legacy of our action: whatever we think, say, and do *counts*.

Contemplating the end of our civilization is not easy because we have a tendency to think of our civilization as a self, as a separate entity, instead of as something impermanent that inter-is with everything else. It may help to visualize the "snowball" Earths of the past, or the tropical Earth of the dinosaurs. We know our planet has had many manifestations. And yet, how can we accept that humanity might fail—that, *if we continue going in the direction we've been going*, there's *no doubt* we will fail? How can we accept something when we want to change it? There's something in that that may feel defeatist. And so, we refuse to accept failure; we refuse to give in to what seems like negative thinking. It's one thing to read it on the page; it's another to truly swallow that stark truth into the depth of our being and make peace with it. Perhaps we don't even dare to try.

Thay is daring us to try. He says, "When you can face the truth and fully accept reality as it is, you will have a breakthrough and be able to have peace." And, with that sobering peace, freedom and clarity will naturally arise, and we'll have the energy we need to do what we can to change the situation. When peace and acceptance is born in our hearts, *we realize we have nothing whatsoever to lose*; we're inspired to do whatever we can to help. In the light of the teachings on karma, whatever we think, say, and do in this moment will make a difference to the next moment, to this evening, to tomorrow, to the next generation.

It's a practice, a training, to keep these radical insights alive in our daily actions, especially in those moments of loss, disappointment, or crisis, when the going gets tough and our feelings get strong. We need the energies of both understanding and compassion to sustain us and our actions. And so the next problem becomes, How can we generate the love that we need? And how can we not burn out while giving?

 *t.d.

Your Deepest Need

Even if we want to help the planet and work for justice, human rights, and peace, we may not be able to contribute anything if we haven't yet been able to fulfill our own most basic needs. Our deepest need is not only to have food to eat, a house to live in, and a partner to love. I've seen many people who have all these things and yet they continue to suffer deeply. There are those who are wealthy who continue to suffer; those who are powerful or famous still suffer deeply. We need something more than these material things.

We need love. We need understanding.

We may have the impression that no one understands us. We think, "If only *one* person could understand me, I'd feel better." But, so far, we have not found anyone who can truly understand our suffering, our difficulties, or our dreams. It is understanding that we need the most—and love.

We also need peace, some deep peace inside. Without that, we're lost. Once we have peace, we are clear enough and calm enough to see the way forward. To have peace inside is a very basic need. Without it, you can't do anything to help others.

And so, we all need peace, understanding, and love, but it seems these things are very rare. You can't get them in the supermarket; you can't get them online. The question is, "How can I myself create the energy of peace, of understanding, and of love?" Meditation is for this. It's a very urgent task. We can learn *how to cultivate*, in any situation, a feeling of peace, understanding, and compassion.

Love starts with observing body and mind. All of us have some kind of suffering, some pain, in our body and in our mind, and so love is needed right away. There may be suffering in the body, or in the mind, perhaps a block of suffering that has been there a very long time—whether it has been transmitted to us by our parents and ancestors or accumulated in our own lifetime. We have to be able to recognize our suffering, and learn how to transform it, so we don't transmit it to future generations.

We can learn a lot from our suffering, and there is always something we can do to transform it into joy, into happiness, into love. It is only by having the courage to encounter our own suffering that we can generate the clarity and compassion we need to serve the world.

Let in the Light

A yogi, a practitioner, is an artist who knows how to handle their fear and other kinds of painful feeling or emotion. They do not feel they are a victim because they know there is something they can do.

You listen to the suffering in you and get in touch with it. Breathe in and out deeply to see, "Why am I suffering? Where has it come from?" Your suffering, your fear, may reflect the suffering of your parents, your ancestors, and the planet. It also contains the suffering of your time, your community, your society, your nation. It's very important not to cover it up with music, movies, or computer games. To have the courage to go home to yourself, to recognize and hold the suffering inside and look deeply into it, may be the most important thing for you to do as a meditator.

The meditator breathes in, and says, "Hello, my fear, my anger, my despair. I will take good care of you." The moment you recognize the feeling and smile to it with love and care, embracing the fear with mindfulness, it will begin to change. That is the miracle of mindfulness. It's like morning sunlight shining on a lotus flower. The bud has not opened yet, but as the sunlight pours down, the photons penetrate into the bud, and after one or two hours of being penetrated by the light, the flower opens itself.

We have the energy of mindfulness generated by mindful walking, sitting, or breathing. With that energy we embrace our fear as gently as the light embracing the flower. When the two kinds of energies encounter each other, there will be a change, a transformation. The energy of tenderness penetrates the fear, anger, or despair. You hold it as dearly as you might hold a wounded child.

If the emotion becomes very strong, you'll feel it coming up. The way to handle it is to put yourself in a stable position and use your in-breath and out-breath to make you solid so it can't sweep you away. Lying down or sitting, you focus your mind on what is called the *dan tian* point, an inch or so below your belly button, and you can even place your hands there. You concentrate 100 percent on your in-breath and out-breath and the rising and falling of your abdomen, and you'll be able to stop the thinking. In that moment, it's very important to stop the thinking because, the more you think, the more your despair and fear will carry you away. Don't be afraid. The wave of emotion is like a storm, and it will go away after some time. You may breathe in for a count of six, seven, eight, or even ten seconds, and you breathe out for ten, twelve, fifteen seconds or more, without thinking. You will feel relief.

Zen in a Storm

In 1976, together with my colleagues and friends in the Buddhist peace movement, we organized relief work to help save the lives of refugees who were escaping Vietnam. In Singapore, we secretly rented three large boats to rescue people adrift on the high seas and quietly took them to other countries for asylum. At that time, the authorities were leaving the refugees at sea to die, sometimes even pushing the small boats back out. And so, if we wanted to help the boat people, we had no choice but to break the law. In one mission, we rescued nearly eight hundred people in the Gulf of Siam, but the Malaysian government refused to allow our boats to enter Malaysian waters. During those days we practiced sitting meditation and walking meditation and ate our meals in silence and concentration. We knew that, without that kind of discipline, our work would fail. The lives of many people depended on our practice of mindfulness.

But, while we were in the middle of trying to find a way to bring the refugees safely to shore in a country that would accept them, our relief program was exposed. The Singapore police came to our door at 2:00 a.m., confiscated my travel documents, and ordered us to leave the country within twenty-four hours. We still had hundreds of people aboard the boats, not yet brought to safety and without enough food or water. Their lives depended on us. There were high winds and rough seas, and one of the boat's engines had broken. What could we do?

I had to breathe deeply. It was an extremely difficult situation. There were more problems than it seemed possible to solve in the twenty-four hours before I had to leave. I realized that I needed to put into practice the words "If you want peace, you have peace right away." You have to *want* it enough. I saw that, if I couldn't have peace in that moment, I would *never* be able to have peace. Peace can be found in the midst of danger. I will never forget every second of sitting meditation, every breath, and every step I took in mindfulness through that night.

At around four in the morning I finally got the insight that we could appeal to the French ambassador, who had been silently supporting us, to intervene in our favor and ask the Singapore authorities to grant us leave to stay just ten more days. Ten days might be just enough to get the people to safety. The ambassador agreed, and at the last minute, we got approval from the Immigration Office to stay. If we hadn't had the practice of meditation—of mindful breathing and mindful walking—we would easily have been overwhelmed by suffering and unable to keep going. We were eventually able to get supplies to the boats and, although it took months to get them ashore and many years for their asylum claims to be processed in refugee camps, their lives were saved.

No Mud, No Lotus

There is a deep connection between suffering and happiness; it's like the connection between the mud and a lotus flower. When you take time to listen to your suffering and look deeply into its true nature, understanding will arise; when understanding arises, compassion is born. We can call this "the mechanics of compassion." You make good use of the suffering to create something more positive: compassion. Just like when you use mud to grow lotus flowers, if there's no mud, there's no lotus. And, in the same way, if there's no suffering, there can be no happiness, no compassion.

There are two mistaken views we tend to have about suffering. The first is to think that when we suffer there is *only* suffering, that all of life is suffering and misery. The second is to believe that *only* when we remove *all* suffering can we be happy. This is also not true. There may be many things we're not happy about, but, *at the same time*, there are many conditions of happiness. To enjoy sitting meditation, for example, doesn't mean that you have to be totally empty of suffering. Each of us has some suffering, but we know the art of handling it.

Don't think that we can only be happy when we remove 100 percent of our ill-being. That is impossible. You may think that if you become a buddha you don't need to practice anymore because a buddha has awakening, insight, joy, and happiness. But awakening, insight, and happiness are all impermanent. If a buddha wants to nourish these things they have to continue to practice making use of suffering, just as a lotus flower that wants to keep blooming has to keep standing in the mud that nourishes it. Similarly, getting in touch with our suffering, embracing and transforming it, generates awakening, insight, and compassion.

Once a disciple asked his teacher, "Where should I look for nirvana?" The Zen master replied, "Right in the heart of samsara!" We have to use the suffering—our fear, our despair, our anxiety—to create happiness, awakening, and insight. The practice is *to make good use* of suffering in order to create happiness. There cannot be one without the other. Suffering and happiness inter-are. We have to find a way to face our suffering and transform it into happiness and compassion, just as we use the place where we fall to help us stand up.

How can we see the good when everything looks so bad? This is a very difficult question. But I believe that, if we have enough time, we can see the positive aspects of something. If you can't yet see it, it's because you haven't yet taken time to look deeply. The practice of meditation is the practice of taking enough time to sit and look deeply to understand.

When I left Vietnam in 1966 to call for peace, I intended to be away for only three months and then go home. All my friends were in Vietnam; my work was there. Everything I wanted to do and everyone I wanted to be with were in Vietnam. But I was exiled for daring to call for peace. And it was very difficult. Although I was already forty, and a teacher with many students, I still had not found my true home. Intellectually, I had a lot of Buddhism and Buddhist training, and could give good talks on meditation, but I had not yet truly arrived. And deep in my heart I wanted to go home. Why would I want to stay in Europe and America? During the day I was very busy, giving talks, attending press conferences, responding to interviews. But, at night, I would see myself going home. I dreamed of climbing a beautiful hill, very green, with beautiful little huts. It was always the same hill. And always, halfway up the hill, I would wake up and remember I was in exile. I tried to train myself to see that Europe was beautiful too; the trees, the rivers, and the sky there were all also beautiful. During the day it was okay. But at night, the dream kept coming back.

Then one day, a few years later, I realized that the dream had stopped. Over time, the sorrow and longing deep in my consciousness had been embraced by concentration and insight. The desire to go home was still there, but I didn't suffer anymore. And, finally, the day came when I felt that, even if I could never go home during this life, it would be okay. There was total emancipation, no more regret. I realized that over there in Vietnam is over here, and over here is over there. Things inter-are. And if I can live deeply here, I am living deeply there at the same time. That insight took more than thirty years to manifest. And so, there are some things we can transform quickly, and others we have to be more patient with. But emancipation is possible. We just need to know the way, the path. Once we see the path, we already begin to suffer less.

The Vietnam War was a bad thing and being in exile for thirty years was a bad thing, but because of that, I have been able to share the practice of mindfulness in the West. Because of that we have Plum Village Practice Center in southwest France and many other practice centers and communities of mindful living in Europe and the US. And so, if we take enough time to look deeply, we see that even so-called "bad" things can produce "good" things, just like the mud that produces lotus flowers. We know that mud is crucial to produce the lotus, but we also know that too much mud will harm the lotus. So, everyone needs a certain dose of suffering in order to grow, just enough for us to learn from. We already have more than enough suffering; we don't need to create any more. The practice of meditation is to look deeply to recognize the suffering and try to understand it. And, when we understand the roots of suffering, the path of transformation and healing will reveal itself.

Are We Done Yet?
The Compost Keeps Coming!

—T.D.

What happens if we *can't* find the lotus, and all we see is mud? One day I was having tea with Thay, and I asked this question. At that time, all I was seeing was mud. "Then you need to look more deeply. The lotus *is* there," Thay said, and smiled, knowingly. I felt a little frustrated. But, over time, I realized that only *we* can find our own lotuses from our mud; no one else can do it for us.

First I had to train myself in how to breathe. It sounds simple and yet it's an incredibly powerful practice. When I came to Plum Village I learned about mindful "belly breathing," and I remember hearing Thay speak about how he could still recall every breath and every step from that vivid night in Singapore. I trained in the basics of mindful breathing and hoped that, if ever I found myself in a crisis, I'd be able to breathe like that. Then, one day, shortly after I was ordained, I heard someone speak about the power of taking ten mindful in-breaths and out-breaths with strictly no thinking. They said that simple practice had changed their life. The trick is that as soon as you have a thought you have to start counting again at zero. "Sounds simple enough," I figured. We have sitting meditation twice a day, an hour for walking meditation, and three silent meals—surely I can do it.

I tried and soon discovered that it's much harder than it seems. It took me more than two months to train myself to breathe ten breaths without a thought. I did so by becoming completely fascinated by the experience of the mechanism of my breathing, from within the body, and by bringing all my attention away from my mind and into the felt experience of the body breathing. Is my breath jagged or smooth, shallow or deep, long or short, irregular or rhythmic? How is my body experiencing movement, temperature, or physical pain? Neuroscientists call this awareness "interoception"; in Buddhism, it's a particular concentration called "mindfulness of the body *in* the body."

I began to train myself to apply this practice in difficult situations—those acute moments in my daily life when something went wrong, or I got hurt, or I was faced with a dilemma that seemed impossible to solve. I learned to step away and take a few immersive breaths, fully aware of the whole experience of each breath. This practice started to give me a safe foundation to be with painful feelings, and gave me space to respond in better ways rather than worse ones. Sometimes it's harder to do it than at other times. I tell myself that if I'm finding it really hard to stop and breathe, if I just can't step back and stop, it's because my ancestors found it difficult, and my society finds it difficult, and my habits find it difficult. I inter-am with all of these conditions. In which case, I tell myself, "Stopping must be the action of heroes!" And that gives me extra determination to keep trying. When I manage in these moments to stop and breathe, each time it feels like a victory, a turning point, a new departure.

I vividly remember one evening in the monastery, an old reaction of overwhelming despair was triggered unexpectedly. My

mind was in chaos and I was flooded with tears. Walking didn't help. Then, through the storm an inner voice came from somewhere deep inside: "Who are *you* to know what to do if you can't even take ten mindful breaths!" So I set about taking my ten breaths, which, having trained myself to do, I figured might take a few minutes at most.

Lying on my back, hands on my belly, drawing my concentration to my breathing, the strength of mindfulness required felt like I was tethering not just one wild horse but a hundred. I counted along my fingers and had to start at zero again and again and again. Finally, through sheer stubbornness I was able to do it: ten breaths without thinking. I sat up drained, relieved, and vividly present. It had taken almost an hour. My next thought was, "Right, the problem. What was the problem again?" To my surprise, the whole landscape of my perception had changed. The situation looked and felt completely different—and all the possible solutions were there, clear as day. I was shocked. It was the first time I realized that there can be moments when it's far better to trust my breathing than to trust my mind.

Like most of us, I didn't learn at school how to handle strong or scary emotions. Instead, I grew up in a society that offers sophisticated and addictive ways to manage the pain and cover it up. Screens offer a thousand worlds to escape into. Some of us may even have found ways to project our own pain onto the world outside us and try to solve it out there. But that's not a solution either.

The challenge is to invest in actively training ourselves to lean inwards and accompany painful feelings when they arise. Fear, despair, grief, and anxiety can be triggered by both internal and external causes, whether it is our own life circumstance and the systems we are a part of, or the injustice, inequality, and destruction we witness across the world. New words are appearing to help us recognize our sorrow and fear for the planet, including "ecological grief," "climate anxiety," "apex guilt," and "solastalgia"—the sadness we feel at the loss or damage of a beloved place in nature.

As a meditator, our task is first of all to take care of these feelings as they manifest in our own body and mind. Trusting in our mindful breathing can give us stability to be with grief and pain in an embodied way. The flow of our in-breath and out-breath is as vivid a measure of our feelings as a seismograph. Whether our breathing is long or short, jagged or smooth, regular or erratic, is a reflection of the feelings that are alive in our heart. And so being 100 percent with our breathing is a way to be with and embrace painful feelings at the base, beyond thoughts, words, and stories.

Taking refuge in our breathing, we allow the feeling to arise, stay for a while, evolve, and pass. We don't deny it or try to change it in any way. The energy of mindfulness allows us to be gentle, compassionate, and curious about what our grief or sorrow is telling us. Being present, and calmly embracing the feeling in the body, gradually gives rise to growing compassion, as well as clarity and courage about what we can do and how we can respond. Since my vivid experience that night in the monastery, when for the first time I could breathe through acute despair, I have learned never to underestimate the power of mindful breathing.

In our society there are those who can't breathe. As a meditator we need to be aware of that fact. There are those who can't breathe because they are sick, and those who can't breathe because the air is polluted, and there are also those whose breath is cut off by the violent injustice of systemic racism and inequality. This is an immense source of suffering for humanity and for the Earth. There is a profound connection between how we treat one another and how we treat the Earth.

Racial justice and environmental justice inter-are. Across the planet, the communities least responsible for climate change and environmental degradation are the ones most disproportionately impacted. There is a profound connection between the way we are harming and exploiting one another, and the way we are harming and exploiting the Earth. Whether we're a victim of injustice, racism, or inequity, or a white person privileged by the system, we are called on to look deeply as part of our meditation. In our sitting and walking meditation, our task is to bear witness to the suffering, touch compassion, and actively explore how we can contribute to healing and change.

———————————✳t.d.

A Dose of Bitter Melon

The desire to practice to transform our own suffering and the suffering of the world is called *bodhicitta*. Sometimes this word is translated as "the beginner's mind" or "the mind of love." It is a powerful source of energy on our path. It nourishes you and gives you the energy you need to overcome the difficulties you encounter. I have had the happiness of keeping my bodhicitta alive for a long time. It's not that I didn't encounter obstacles; I encountered a lot of obstacles. But I never gave up because the energy of bodhicitta was very strong. You should know that, as long as the beginner's mind is still alive and powerful in you, you don't have to worry. You can be of service to the world your whole life. You'll be happy, and you'll bring others a lot of happiness too.

In Asia, there is a marrow-like vegetable known as bitter melon. The Vietnamese word for "bitter" also means "suffering." Those who aren't used to eating it have a hard time; it is very bitter. But in Chinese medicine we believe that bitterness is good for your health. Bitter melon is refreshing and cooling. It may be bitter, but it's delicious, and we know it's doing us good.

We have a tendency to avoid what is bitter and to run away from our suffering. We're not aware of the goodness of suffering, the healing nature of suffering. Some of us practice sitting meditation in order to escape suffering. It gives us some stillness and relaxation and helps us leave behind our difficulties and disputes, and we feel a little bit of peace and happiness. But that's not the true purpose of meditation.

Zen master Linji shouted at his students, "Don't sit like that! Don't sit like a rabbit retreating into its hole!" Sitting is not for avoiding suffering.

In meditation we have stillness, we have relaxation, we have mindfulness, concentration, and insight. And we also have joy and happiness, cultivated by letting go, by leaving behind our daily worries and concerns and craving. Letting go is the first step.

Even so, there may still be a mental discourse going on in your mind. And so, you need to stop the mental discourse to cultivate stillness and enjoy some peace and happiness. You are sitting there, enjoying your breathing, enjoying the stillness, enjoying the quiet inside.

And yet, even that is not enough. You have to go deeper. On the surface there may be stillness, but underneath there are hidden waves. Sitting meditation is to use your intelligence and concentration to go deeper to transform the pain, fear, anxiety, and suffering that is there in the depths of your consciousness.

The suffering that is there may be suffering you endured as a child or it may be the suffering of your parents or grandparents that they were not able to handle and that they transmitted to you. Even if you have only a vague feeling of your suffering, you have to practice getting in touch with it and use your insight to recognize it. It's like eating bitter melon. You're not afraid. You know that bitter melon is helping you.

When suffering is emerging, adopt another attitude. Don't try to run away. This is my recommendation. Stay where you are and welcome it, whether it is anger or frustration or a longing for something that is not satisfied. Be ready to say hello to it, be ready to embrace it tenderly and live with it. And you will discover, as I have, that when you can accept it and welcome it, it does not bother you anymore. Just like the bitter melon, it is healing you. If you do not accept your suffering, if you don't embrace it tenderly, you won't know what it is. That suffering can teach us and, when we understand it, it can bring us joy and happiness. Allow yourself to suffer a little bit. Many children hate bitter melon at the beginning, but, when they grow up, they love to eat bitter melon soup.

The most difficult case is when you know the suffering is there but you can't name it; it's too vague. It's there in you, it's real, but it's hard to identify. There's some blockage, some resistance in our consciousness. Every time you're about to touch it, you avoid it. This may have been happening for a long time, which is why you haven't had a chance to identify it clearly. And so, you must resolve to not continue like that. Every time it comes up, welcome it. And, with the energy of mindfulness, stay in the present moment, with vigilance, and you'll be able to identify it.

Meditation is centered in the present moment. You don't have to go anywhere or travel back in time to your past or childhood to meet your suffering and its roots. You just stay right in the present moment and observe.

Reclaim Your Agency

Zen master Linji declared that we must each be our own master, and not be a victim of our surroundings. We must keep our freedom, even when things around us do not go as we wish. It is our responsibility to be master of the situation and to make use of whatever situation we are in to awaken. Wherever you are, you can be sovereign of yourself. An active person always asks, "What can I do, what can we do, to keep this situation from getting worse and to help it improve? How can I help the other person or people to change?"

Giving rise to that mind of love, and the vow to help, we are no longer passive, no longer the victim. We become active again. Bodhicitta gives us energy and the will to be active and to change. It's very important. Even though we haven't done anything yet, just that insight and the will to change already lessens our suffering 80 to 90 percent.

If we look carefully, we see that there is no one in the world who has not been the victim of difficult situations. Society is full of discrimination, violence, inequality, hatred, craving, and greed. People are overwhelmed by these things and make each other suffer and make other species and the planet suffer. So, we can't say there's anyone who's not a victim of something. And we should remember that, even in ourselves, we also have the seeds of discrimination, anger, craving, violence, and unskillfulness. When you are able to transform yourself, you're in a situation to help transform those you believe to be your oppressor and the source of your suffering. This has been my own experience and practice. I do not have enemies, even though I have experienced a lot of suffering, a lot of injustice. There are those who have tried to kill me, to suppress me, but I don't see them as my enemies. I want to help them. I have changed and transformed myself, and that is why I no longer see myself as a victim.

When we get angry, the seed of anger in the depths of our consciousness comes up, and our mind tells us that we're suffering because of that person or that situation. But, as soon as we practice conscious breathing, and recognizing and embracing our anger, our mind begins to restore its sovereignty and can declare, "I don't want to be a victim of my anger. I want to be myself. I want to initiate change." In this way, mindful breathing becomes a kind of practice that restores your sovereignty and develops your free will.

With the energies of understanding, insight, and compassion, you are free, and you can help others become free. Practicing in this way you will be able to transform your heart and mind, and you become a bodhisattva. You will be in a position to help those who discriminate against you, those who suppress you, or those who try to kill you.

Every time we fall is an opportunity to stand up again—that's the attitude of one who is active: *every time I fall, I'll stand up again so life can get better*. That's the attitude: although there are obstacles and challenges, we don't allow ourselves to be overwhelmed by them. We stand up like a hero. With that intention, a great deal of suffering falls away already.

In the Company of Bodhisattvas

A bodhisattva is a living being (*sattva*) who has woken up (*bodhi*). Anyone with happiness, mindfulness, peace, understanding, and love can be called a bodhisattva. Bodhisattvas have a deep aspiration to help others, and that aspiration is a tremendous source of energy and vitality.

Bodhisattvas aren't only human. A deer, a monkey, a mango tree, or a rock can all be called bodhisattvas because they are offering freshness, beauty, and refuge to the world. Bodhisattvas aren't great beings up in the clouds, or statues of metal or wood on an altar; they are all around us. The pine in the front yard can be a bodhisattva offering peace, oxygen, and joy. The Diamond Sutra reminds us that, if we're caught in the sign "human," we won't be able to see the buddhas and bodhisattvas all around.

In Buddhism we consider the planet to be a bodhisattva: a real, authentic, great bodhisattva. One of Earth's many names is "Great Refreshing Earth Bodhisattva." Our planet is the most beautiful of all bodhisattvas. She has the qualities of endurance, solidity, creativity, and non-discrimination; she embraces and sustains everyone and everything. This is not to say that Earth is inhabited by a spirit or to say that there is a spirit somehow behind or within the planet. We shouldn't get caught in thinking there can be "matter" inhabited by "spirit." The Earth cannot be described in terms of matter or mind; the planet transcends both categories. The great Earth is not without perceptions, without feelings; it's not impersonal.

How could mere matter do the wonderful things the Earth does? The great Earth is not human but has given rise to humans, including exceptional humans of great compassion and understanding. The Earth is nothing less than the mother of all buddhas, bodhisattvas, saints, and prophets.

The Earth is a bodhisattva we can take refuge in, not only outside of us but within us. We don't need to die to return to the Earth; we are already here within the Earth. Whether walking or sitting, we can breathe with the Earth and learn to allow the Earth to be in us and around us. The Earth has the power to heal herself and to heal us, and we can trust in that power. To see this is not blind faith but can come from observation and experience; it's not something others tell you to believe in. Every time you suffer, every time you feel lost or alienated, you can get in touch with the Earth to restore yourself. By connecting with the Earth within ourselves, healing begins to take place; healing becomes possible.

We don't need to go somewhere to find bodhisattvas. We don't need to get on a plane or travel anywhere, not even to the four great sacred mountains of the bodhisattvas in China. Zen master Linji said it is possible to sit right where you are and get in touch with them; they're right here in the very present moment, in our own hearts and minds. Why look for them outside of us?

In the Buddhist tradition, we speak of the bodhisattva Manjushri, the bodhisattva of great understanding. Inside us, too, there's the seed of understanding. Samantabhadra is the bodhisattva of great action; we, too, have the seed of action. Avalokiteshvara is the bodhisattva of great compassion; we, too, have the seed of compassion. We only need to water these seeds in our consciousness, and we are in touch with the bodhisattvas right in this moment. What we're looking for is both outside and inside and is neither outside nor inside. Outside of what? Inside of what? According to the Diamond Sutra dialectics—"A is not A, and that is why it can truly be called A"—a bodhisattva is not a bodhisattva, and that is why they can truly be called a bodhisattva, because a bodhisattva is made *only* of non-bodhisattva elements.

Bodhisattvas are living beings with the qualities of understanding, compassion, action, reverence, and so on—and you also have these qualities. You don't need anyone else to confirm this; you know it for yourself. The training is to keep these qualities alive in you. Whenever we are caught in despair, for example, and we feel resentment and hatred toward people or situations, we know we don't have understanding yet. But, as soon as we look more deeply and the seed of understanding arises in us, light shines and the darkness disappears. That is Manjushri, the bodhisattva of understanding, really existing in you.

To become a bodhisattva is possible. Bodhisattvas are not people who don't have difficulties. Difficult moments may come, but bodhisattvas are not afraid because they know how to handle them. Every one of us has to find our own lamp, our own light, and offer it to the world. Mindfulness is a kind of light, an energy, that helps us know where we are, what is happening, and helps us know what to do and what not to do for peace, compassion, and happiness to be possible. We know this is a very important moment to be alive on this beautiful planet. With the light of mindfulness we become a bodhisattva; we provide the world with light. You let awakening shine around you. You look with the eyes of the bodhisattvas; you act with their hands. And, if we do that, there is no reason to be pessimistic about the situation of the world. A bodhisattva with such deep intention is free from despair and can take action in peace and freedom.

Engaged Action

We coined the term "Engaged Buddhism" in the 1960s when the war was very intense in Vietnam. We practiced sitting meditation and walking meditation, but we could hear the bombs falling outside and the cries of the wounded. To meditate is to be aware of what is going on, and what was going on at that time was suffering and the destruction of life.

Once you know what is going on, you're motivated by a desire to *do* something to relieve the suffering—both in you and around you. And so, we had to find a way to practice mindful breathing and do walking meditation *while* helping those wounded by the bombs—because, if you don't maintain a spiritual practice during the time you serve, you will lose yourself and you will burn out. And so, we learned to breathe, to walk, to release the tension so we could keep going. These are the origins of Engaged Buddhism; it was born in a difficult situation, where we wanted to maintain our practice while responding to suffering. In such a situation, *anything* you do in mindfulness— whether it is social action, drinking your tea, sitting in meditation, or making breakfast—you do it not only for yourself. You are preserving yourself so you can help the world.

This is the attitude of bodhisattvas: to practice meditation not only for yourself, but for the world, to relieve the suffering. And, when others suffer less, you suffer less. When you suffer less, they suffer less. That is interbeing. There is no separation between yourself and others. You do not live just for yourself; you live for other people. Your peace, freedom, and joy also profit others; you are already helpful. And so, when you breathe mindfully or walk mindfully and create joy and peace, that is already a gift for the world. Lighting up the energy of mindfulness in the heart of your family, your community, your city, or your society *is* engaged action. Compassion and peace radiate from your person.

Look at the tree in the yard. Every tree should be a real tree: stable, serene, and fresh. If a tree is a healthy, beautiful tree, everyone profits. But, if the trees are anything less than trees, we'll all be in trouble. That is why, if we know how to be a healthy human being, happy and compassionate, we are already serving the world. Wherever we are, we are helpful. And so, it's not a question of having to choose between helping others and helping ourselves. Master Tang Hoi said, "An arhat is a bodhisattva, and a bodhisattva is an arhat." It means, the one who has awakened is a bodhisattva. There is no more discrimination, no artificial boundary between yourself and others. It always begins with you. And you don't need to wait ten or twenty years to be helpful and have an impact; right away you can help many people.

We know that we have the "ultimate dimension" where you don't have to do anything. It's very nice to dwell in the ultimate dimension. We should all learn how to do it. And then you have the historical dimension, where there is suffering, injustice, inequality, exploitation, and so on. The question is, when we suffer in the historical dimension, how can we touch the ultimate dimension so we can stop suffering from fear, despair, and loneliness? How can we bring the ultimate dimension to the historical dimension?

I propose that we need another dimension, the action dimension. The action dimension is the realm of the bodhisattvas, the kind of energy that helps us bring the ultimate into the historical so we can live our life of action in a relaxing and joyful way, free from fear, free from stress, free from despair. Every one of us should be a bodhisattva, bringing the ultimate dimension into the present moment so we can arrive and stop running, so we can be relaxed and joyful, so we can make peace and enjoyment possible for humankind and for other species on Earth.

true
love
heals

PART 2

THE ACTION DIMENSION
A NEW WAY
OF LIVING

The image we have of an enlightened person is someone with freedom and spiritual strength who is not a victim of their environment. An enlightened person sees themselves clearly, knows who they are, and has a clear understanding of reality—both their own nature and the reality of society. This understanding is the most precious gift that Zen can offer.

The enlightened person's way of being is the most fundamental positive contribution Zen makes to the world. Zen is a living tradition, one that can train people in this understanding and in a way of life that is healthy, resilient, and balanced. The art and thought that stem from the insights of Zen will also have these qualities of vitality, stability, and peace.

Whether or not we can cultivate these qualities of ease and freedom is a question of awakening. The world doesn't need another ideology or doctrine but the kind of awakening that can restore our spiritual strength. With real awakening we can see the situation clearly and reclaim, from the social and economic systems of our own creation, our sovereignty as human beings. The way out is to engage in a new way of living that can restore our sovereignty and our humanity.

Awakened Action

Action should be based on the foundation of being. If you don't have enough peace, understanding, and tolerance, or if you're burdened by anger and anxiety, your action will have little value. And so, the quality of action depends on the quality of being. In Zen we speak of the action of non-action. There are those who don't seem to be doing very much, but their presence is very crucial for the well-being of the world. And there are others who keep trying to do things but, the more we do, the more troubled society becomes because the foundation of our being is not good enough. Sometimes you don't do anything, but you do a lot. And sometimes you do a lot, but you don't do anything; it doesn't help. There are even those who meditate a lot, but their anger and jealousy just remain the same.

In the historical dimension, yes, there are things we need to do, actions we need to take to save, to nurture, to heal, to reconcile. But, in the ultimate dimension, you can do everything relaxingly and joyfully, with no worries at all. That is what it means to be "acting the non-acting action." You are very active, but you're so relaxed you seem to be doing nothing at all. You enjoy every moment because you're acting from a base of non-action, without striving or rushing.

In this way of acting, our actions become a true expression of our love, care, and awakening. It's not that *we* have to take action. If we're awakened, *action will naturally take us*. We can't avoid it.

In the meditation tradition we speak of the ideal of the "businessless" person: someone who is free, at ease, and no longer striving or seeking. The Vietnamese word is *người vô sự* and the Chinese is 無事人. A free, businessless person is very active in helping the world, and in helping to relieve suffering, but is never carried away by their surroundings or the work they are doing. They don't lose themselves in their ideals or projects. This is very important. We shouldn't do things for praise, fame, or financial profit. We shouldn't do things to run away or avoid something else. We act out of love.

When we do something out of love, we can feel the happiness it brings us. But when we do something *without* love, we suffer. We find ourselves saying, "Why do I have to do this on my own? Why aren't others helping?" The important thing is to not lose ourselves in our action. We remain sovereign in every situation. We are at ease in ourselves, and we are free.

The Meditator, the Artist, and the Warrior

In every one of us there is a meditator, a yogi. That is the wish to meditate, to practice, to become a better person, to bring out the best of ourselves, to get enlightened. Our inner meditator brings us lucidity, calm, and deep insight. That is the buddha-nature in us. We may want to become better people, and yet there are times we don't practice, we don't train, not because we don't want to but because we haven't yet created the right conditions.

In each one of us there is also an artist. The artist is very important. The artist can bring freshness, joy, and meaning to life. You need to allow the artist in you to be creative so you can always feel and enjoy the nourishment in your practice of mindfulness. Many of us can't stand monotony. If we have too much of something, we want to change it, even if we know it's good. This is only natural.

You may ask, "How can we keep going on a path we want to go on, and keep going to the end?" Of course, you need patience. But you also need something else: the path should be joyful, nourishing, and healing. So, we have to find a way to *create* that joy every day.

We have to organize our daily life so it's not repetitive and so each moment can be a *new* moment. We must find creative ways to keep our bodhicitta, our beginner's mind, alive and nourished.

Whether you're eating in mindfulness, driving in mindfulness, or practicing walking meditation or sitting meditation, you have to invent new ways of doing it, so that the breathing, walking, and sitting always brings you delight, solidity, and peace. On the outside, it may look the same, but you're walking as a new person, you're sitting very differently; you are *evolving.* I can tell you I never get bored of walking in mindfulness. When I walk, every step is a delight, and not because I am diligent or disciplined but because I allow the artist in me to operate and to make my practice new, interesting, nourishing, and healing.

Practicing mindfulness can *always* be healing and nourishing, if we know how to be creative. We shouldn't practice like a machine but as a living being. According to Master Linji, if, while walking or eating or going about your day, you can create even just one flash of mindfulness, that's good enough. Just 1 percent success is good enough because that 1 percent can be the ground of many other percents.

In every one of us there is also a warrior. The warrior brings a determination to go ahead. You refuse to give up. You want to win. And, as a practitioner, you have to allow this fighter in you to be active. You don't become a victim of anything. You fight in order to renew your meditation practice. You fight in order not to allow things to become boring. And so, the meditator goes together with the warrior. We should not be afraid of obstacles on our path. In fact, there are many things that can discourage you. But, if your energy of bodhicitta is strong, if your warrior is strong, you can overcome these obstacles, and every time you overcome them, your bodhicitta will get stronger. In this way, obstacles are not really obstacles. They are an accelerator of wisdom, of aspiration.

The meditator, the artist, and the warrior are not three separate people; they are three aspects of your person. And you should allow all three aspects to be active at the same time in order to have balance. We have to mobilize them all and never let one of them die or become too weak. If you are an activist, a political leader, or a leader in your community, you have to know how to cultivate these three aspects within yourself so you can offer balance, steadiness, strength, and freshness for those around you.

Love in Action

—T.D.

Sister Chan Khong, Thay's most senior disciple, is a bright light in our Plum Village community. Since she first began assisting Thay in the 1960s, she has been a pioneer of Engaged Buddhism and true bodhisattva. She is known as "Sister True Emptiness" (and as "Sister Bare Feet" to the children), and her life is a testament to the power of compassion to heal, protect, and save. In her teens she created a program to help poor children in Saigon's slums, before meeting Thay and helping start his engaged social work programs. Her spiritual practice sustained her as she delivered aid under gunfire and led teams to bury the dead after the terrible massacres of 1968. Against the backdrop of war she completed doctoral studies in biology, and joined the Buddhist Peace Delegation at the Paris Peace Talks. It was Sister Chan Khong who helped Thay convene one of the first scientific conferences on the environment and, in the early 1980s, helped find derelict farms in the French countryside where he could establish a community for mindful living.

I was twenty-one when I first met Sister Chan Khong in Plum Village, during one of her remarkable sessions of deep relaxation and "Touching the Earth"—a powerful meditation practice in which you contemplate interbeing with ancestors and the Earth, combined with the movement of full-body prostrations. At that time, I had studied at a good university, and had read a lot of books, but being with Sister Chan Khong and the nuns that summer was the first time I encountered living wisdom and wise women. Then in her sixties, Sister Chan Khong had sparkling eyes and a radiant face that traced a life of smiles. As a young person we might hope to meet someone like Sister Chan Khong once in our lifetime. I never imagined I would have a chance to accompany and assist her, and learn from her incredible example.

I vividly remember one trip to the European Parliament in Brussels, Belgium, soon after I ordained. Sister Chan Khong was going to speak out to protect the safety and religious freedom of Thay's young monastics in Vietnam, and Thay assigned a few of us to support and help her. My job was to set up meetings and press interviews, draft briefing texts, and liaise with human rights NGOs. The swords of action took the form of laptops and cell phones. We were staying in a small Tibetan temple in the city center, and Sister Chan Khong delighted in cooking for our team and, with joy, insisted we sing together after dinner. The next morning, we joined the temple's meditation practice and prepared to set off to the parliament on foot. Sister Chan Khong stopped us on the doorstep, and taught us how to walk in silence and concentration. With every step we were to follow our breathing, touch peace and compassion in our hearts, and invite our spiritual ancestors to walk with us.

Sister Chan Khong's meetings with the different representatives for foreign policy, human rights, and religious freedom showed me what love in the action dimension looks like. Sister Chan Khong knew that our success did not depend on how good our documents were or what political support we had. It depended on touching the hearts and humanity of every person we met.

She knew if we could awaken them to the truth that was right before their eyes, they would know exactly how to help. Sister Chan Khong's love and charisma won the hearts of even the most hardened bureaucrats. Sometimes she smiled, sometimes she cried, and sometimes both at once. She knew she was speaking out to defend the right and freedom of the next generation of engaged Buddhists to operate and continue Thay's legacy in Asia. Sister Chan Khong's actions were fueled by love, and sparked love in others. The impact of those meetings continues to resonate today.

Our schedule was relentless: from Brussels we would travel on to Strasbourg, Geneva, and Paris. I remember we took a short break in one of the smart cafes in the European Parliament precinct. Sister Chan Khong walked straight past the table to lie down across the cafe windowsill. She closed her eyes, placed her hands on her belly, and began to practice mindful breathing and deep relaxation. The young waiter smiled, a little taken aback, and then nodded with respect. "But of course. Please, Madame."

Sister Chan Khong is a free spirit and a force of nature. She knows exactly how to balance the concentration, peace, and calm of a meditator with the creativity and joy of an artist, and the strength and endurance of a warrior. Her sharp mind is clear, free, and as quick as lightning. She knows when to engage, and when to let go and move on; she knows exactly which actions will help relieve suffering, and which won't. It is the fruit of years of mindfulness practice, cultivating compassion, and training the mind right in the heart of action.

Training the Mind

Mindfulness, concentration, and insight are three energies that help us generate happiness and handle suffering. We call these energies "the triple trainings" (三學 in Chinese and tam học in Vietnamese). You *train* in mindfulness, you *train* in concentration, and you *train* in insight. Another word for the practice of meditation is bhāvanā. It means to train, to practice, to cultivate. If something is not yet there, we can produce it, just as a farmer cultivates a field to grow wheat or corn. *Bhāva* literally means "being there." So, we produce, we bring something into existence, and that something is joy, peace, and freedom. In English we use the word "practice." If our practice is good and solid, we have no need to be afraid for the future. With our practice we are training ourselves to generate joy, happiness, peace, harmony, and reconciliation, and to handle pain, suffering, separation, and misunderstanding.

The opposite of mindfulness is forgetfulness. Forgetfulness means that you are pulled away by the past, by the future, by your projects, by your anger, by your fear. You're not truly alive. Every one of us has a seed of forgetfulness and a seed of mindfulness inside, and if we train ourselves a little bit— drinking our tea mindfully, breathing in mindfully, or taking a shower mindfully—after a few days, the seed of mindfulness in us will become stronger. With mindfulness you are more concentrated, and with that concentration you'll see things more deeply and clearly. Any decision you make will be wiser, and everything you do will have a better quality. When you're with another person you bring more mindfulness and concentration into the relationship, and the relationship will be deeper. And with training in a daily practice of breathing, walking, and doing things with awareness, the seed of mindfulness in you can grow bigger every day.

Zen Master Linji was very intelligent when he was younger and a diligent student. But in the end, he let go of studying books and texts in order to follow Zen practice. This doesn't mean that we shouldn't study; Master Linji had a firm knowledge of Buddhist teachings. Studies are necessary, but formal study in itself does not lead to transformation and awakening. Many of us are willing to spend six or eight years studying to get a diploma. We believe it's necessary for us to be happy. But very few of us are willing to spend three or six months, or even a year, training ourselves to handle our sadness or our anger, to listen with compassion, and to use loving speech. If you can learn how to transform your anger, sadness, and despair, and if you can learn how to use loving speech and deep listening, you can become a real hero capable of offering happiness to many people.

A Code of Ethics

In our time of globalization, harmony will be impossible without some kind of shared values or "global ethic." The "Five

Mindfulness Trainings" are five short paragraphs that express a Buddhist contribution to a global spirituality and ethic. They propose a spiritual practice that can bring about true happiness and true love, protect life, restore communication, and bring about the healing of the planet and of every one of us on Earth. They are a way out of this difficult situation in the world. The insights of no-self and interbeing are a firm foundation from which you can change your life and behavior. From these insights, right action—for your own well-being and the well-being of our planet—will naturally flow. Following the path of the Five Mindfulness Trainings, you can already set off on the path of transformation and healing and you can become a bodhisattva, helping to protect the beauties of diverse cultures and helping to save the planet.

The Path Is Made by Walking

—T.D.

In Sanskrit, the word for path is *mārga*. It's not some kind of well-worn road but a craggy path winding its way up the mountain. In times like ours, it may be hard to see a path forward; everything is so foggy and uncertain. How can we trust what we see? Which way should we go?

As you will discover in this book, Thay describes mindfulness as a path, not a tool. It's not a tool *to get something*—even if that something is relaxation, concentration, peace, or awakening. It's not a *means* to an end, a means to improve productivity, wealth, or success. In true mindfulness, we arrive at the destination *every step* of the way. That destination is compassion, freedom, awakening, peace, and non-fear. True mindfulness can never be separated from ethics. If the insight you get from mindfulness is real, it will change how you see the world and how you want to live.

To develop the radical insights introduced in Part 1—the insights of interbeing, of the ultimate dimension, and of transforming the bitter melon of suffering into compassion—we need a regular, solid practice of meditation and mindfulness. We need silence, sitting still, and time in nature, but we also need a framework for being mindful in our way of working, consuming, speaking, listening, loving, and interacting with the world. Mindfulness is not only for the quiet of the meditation cushion; it is for the three-dimensional reality of our daily life. It's precisely by applying these insights and teachings against the rough rock-face of reality that we'll be able to walk a path of transformation.

The following five chapters capture Thay's powerful teachings on each of the Five Mindfulness Trainings; five essential precepts that envisage a new way of living on our planet, based on the radical insight of interbeing as taught in the Diamond Sutra. In "Reverence for Life" we explore the ethics of non-violence. In "Deep Simplicity" we re-examine the ideas of happiness that may be destroying our planet and society. In "Right Fuel" we look deeply at what is driving our actions and dreams. In "Brave Dialogue" we

discover new ways of listening and speaking that can support collaboration and inclusivity. And in "True Love" we learn the power of compassion to bring healing and change. Each of these five chapters is accompanied by the text of one of the trainings or "precepts." These ethics may be unfamiliar or unexpected; they may even be challenging. And yet it's only by allowing our ideas and habits to be disrupted that we can find a new way out. If we're going to help our society and planet, we will need to cultivate reverence for life, deep simplicity, real love, and brave dialogue; and we will need to sustain ourselves with the right kind of fuel. Here is an ethical compass, a North Star, to guide our steps.

REVERENCE FOR LIFE: NON-VIOLENCE IS A PATH, NOT A TACTIC

Spiritual Strength

Every one of us should have a spiritual dimension in our life so we can confront and transcend the challenges and difficulties we encounter every day. One way we can speak about spirituality is in terms of *energy*: the energy of awakening, of mindfulness—the kind of energy that helps us to be there, fully present in the here and now, in touch with life and the wonders of life.

We tend to distinguish between the spiritual and the non-spiritual. We associate the spiritual with the spirit and the non-spiritual with the body. But this is a discriminative way of seeing things. Say, for example, I make some tea. We need the tea leaves, the hot water, the teapot, and a cup. Do these elements belong to the realm of the spiritual or the non-spiritual? As soon as I pour the hot water into the teapot with mindfulness, there is the energy of mindfulness and concentration in me, and suddenly the tea, the water, and the teapot all become spiritual. And, as I lift up my cup and hold the tea in my hands with mindfulness and concentration, the act of tea-drinking becomes something *very* spiritual. Anything touched by the energy of mindfulness, concentration, and insight becomes spiritual, including my body enjoying the tea. And so, the distinction between the sacred and the profane is not absolute. We live *in* the world and what is so-called "worldly" can become spiritual once we bring the energy of awakening to it. Mindfulness, concentration, and insight can be generated in every moment of your daily life, and it's these energies that make you spiritual.

When I entered the temple as a very young monk, I was taught to recite these lines the first moment I woke up:

Waking up this morning, I smile
Twenty-four brand-new hours are before me
I vow to live each moment deeply
And to look at all beings with the eyes of compassion

At the time, I didn't realize this poem was deep and meaningful; I didn't understand. Why do I have to smile in the morning when I wake up? Later on, I learned that this smile is already a smile of enlightenment. As soon as you wake up, you realize that *you have a life*; life is in you, life is around you, and you smile to life. You greet life with a smile so you can really feel alive and feel the energy of being alive in you. You are generating the energy of mindfulness, and that makes you spiritual right away.

As you recite the second line, your smile gets deeper as you realize that you have twenty-four brand-new hours to live, twenty-four hours delivered to your door, to your heart. So, you smile a smile of awakening and joy; you cherish life and resolve to make good use of the hours given you to live.

For Christians, wherever the Holy Spirit is, there is life, there is forgiveness, there is compassion, there is healing. These energies are also generated by mindfulness, so we could say that the Holy Spirit is another word for mindfulness. It is the kind of energy that makes you alive, compassionate, loving, and forgiving. What is important is that energy is in you, and if you know how to cultivate it, you can make it manifest.

And when, with the energy of mindfulness and concentration, we encounter suffering, does that suffering belong to the worldly or to the spiritual? If you allow suffering to overwhelm you, it's not spiritual. But when you know how to recognize and embrace it and look deeply into it so understanding and compassion can arise, even suffering becomes spiritual. That doesn't mean we need to create more suffering; we already have more than enough. But we can *make good use of it* to generate compassion. The suffering may be there in the tension and pain in our body. And when, with the energy of mindful breathing and walking, we release that tension and pain, that is a spiritual practice. And when, with the energy of mindfulness, we embrace painful feelings and emotions like anger, fear, violence, and despair, and bring peace to our body and mind, that, too, is a spiritual practice.

In 1966 I was invited to the US to speak out against the war in Vietnam. There was a growing movement in the US demanding peace and an end to the war, but it was hard to get peace. I recall one talk I gave, when an angry young American stood up and said, "You shouldn't be here! You should be back in Vietnam with a gun, fighting American imperialism there." There was a lot of anger in the peace movement.

My response was "The roots of the war in Vietnam are here in America. The American soldiers in Vietnam are also victims—they are victims of a wrong policy. That is why I have to come here to tell the American people that the war is not helping."

Whatever is happening, we need to keep our understanding and compassion alive, so we don't lose ourselves in anger and hatred. And so, as I toured America and spoke to groups in the peace movement, I shared that, if you have a lot of anger in you, you cannot achieve peace. You have to *be* peace before you can *do* peace. Gradually, together with my friends who were spiritual leaders, we found opportunities to introduce nonviolence and a spiritual dimension to the peace movement.

In difficult situations we all need a spiritual practice in order to survive and to keep our hope and compassion alive.

Every one of us on Earth should bring a spiritual dimension to our daily life, so we will not be swept away and so we can handle our suffering and take care of our happiness. We have to go home to ourselves and look deeply, and that is the work of the spirit. Our century should be a century of spirituality. Whether we can survive or not depends on it.

Do You Love Yourself Yet?

If you don't respect yourself, it will be difficult to love and respect others or the Earth. When you're caught in the idea that *this body* is you, or *this mind* is you, you underestimate your value. But, when you can free yourself from the notion of self and see your body and mind as a stream of being of all your ancestors, you'll begin to treat your body and mind with more respect.

You may feel that you don't deserve love. But everyone needs love—even the Buddha. Without love, we cannot survive. So, we shouldn't discriminate against ourselves; you need love, you deserve love, *everyone* deserves love. Your ancestors in you all need love. So why deprive them of love? They are still alive in every cell of our body. Perhaps in their lifetime they did not get enough love. But now we have a chance to offer them love by loving and taking care of ourselves.

You are one of the wonders of life, even if you believe otherwise—even if you despise yourself or think of yourself as nothing but suffering. The maple tree outside is also a wonder, and so is the orange you are about to peel. And you, who are about to peel the orange and eat it, you are also a wonder. It is only your anger, fear, and complexes that prevent you from seeing it. You are as wondrous as the sunshine and blue sky.

It is possible to train ourselves to be able to breathe in and out mindfully and recognize the many good things that have been transmitted to us: the seeds of compassion, understanding, love, and forgiveness. We can have confidence in ourselves because we can see our ancestors in us. There is democracy today because our ancestors fought hard for it. We have beautiful cities, art, literature, music, philosophy, and wisdom because our ancestors have created them. Your ancestors are there in you. If they can do it, you can do it. You *believe* in yourself, and you trust that you can continue what they could not do in their lifetime.

In Vietnam every home has an altar for our ancestors with images of both spiritual ancestors and blood ancestors. Every day we light a candle and a stick of incense, or offer flowers, and remove the dust. We tell our children about how they lived and the qualities they had and it inspires and unites us. Each one of us can go home to our roots to rediscover the values of our heritage.

Master Linji taught that we have to have confidence in our own seeds of awakening, liberation, and happiness and not go looking for them outside of us. In your body, your mind, your spirit, you have all the elements you need to heal. You already have the elements of awakening, enlightenment, and happiness in you; you just need to come back to yourself and get in touch with them.

Wake Up to the Wonders

I remember a few years ago, when I came to northern California to lead retreats and give talks, I stayed in a little hut in a monastery up on the mountains in a giant sequoia forest. One day, a journalist from the *San Francisco Chronicle* came to interview me about mindfulness. I invited him to have tea before the interview, and we sat outside the hut, at the foot of a stand of huge sequoia trees. I suggested he forget about the interview and just enjoy tea with me. And he was good enough to say yes.

So, I prepared tea, and we enjoyed the fresh air, the sunshine, the sequoia trees, and the tea. I was aware that, in order to write a good article on mindfulness, you have to taste something of it and not just ask questions and get some answers. That wouldn't help readers to understand what mindfulness is. So, I tried to help him to enjoy the tea: mindfulness of drinking tea. Apparently, he enjoyed having tea with me and forgetting all about the interview. After that, we had a good interview, and I walked with him back to the parking lot.

Halfway to the parking lot, I stopped. I invited him to look at the sky, breathe, and smile: "Breathing in, I am aware of the sky above. Breathing out, I smile to the sky." We spent one or two minutes just standing there looking at the sky and breathing and enjoying the blue. When we came to the car, he told me that it was the first time he had seen the sky like that. He really got in touch with the blue sky. Of course, he had seen the sky many times before. But that was the first time he really saw the sky.

Albert Camus, a French novelist, wrote a novel called *L'Étranger*, "The Stranger." The novel is about a young man who committed a crime. He killed someone and is given the death penalty. Three days before the execution, lying down in his prison cell, he looked up. And suddenly he saw the blue sky through a skylight. Of course, he had seen the sky many times, but this was the first time he could see it so deeply—just a small square of sky. Perhaps it was because he was about to be executed that he cherished every moment left for him to live. That awakening, that capacity of getting in touch with reality, is called by Camus in the novel *le moment de conscience*: a moment of consciousness, a moment of mindfulness. Thanks to mindfulness, the young man could get in touch with the blue sky for the first time.

A few moments later, you read that a Catholic priest was coming to offer the young man his last rites. But he didn't want to receive the priest because he was aware that he only had a very, very short time left to live his life and he didn't want to waste it doing something he didn't believe in. He felt he had woken up and that the priest was still in the dark, in forgetfulness, and not fully alive. He used the phrase *il vit comme un mort*: he's living as a dead man.

Many of us are like that: we are alive, and yet we live as dead people because we don't have that moment of awakening in us. We are not mindful that we have a body, which is a wonder. We are not mindful that there are wonders of life all around us, including the blue sky. We walk like sleepwalkers. We are not really alive. It's as though we're switched off; we're not "on" to be there and live our life deeply. We need a kind of waking up in order to be alive again. This is a practice of resurrection. Many of us live as though we're dead, but, as soon as we know how to breathe mindfully and to walk mindfully, we can resurrect ourselves and come back to life.

It was because I had that story in mind that I invited the journalist to stop on the way to the parking lot and look up at the sky, breathe, and get in touch with it. And it was a successful exercise of mindfulness. The article he wrote about mindfulness turned out to be a very good one because he had tasted mindfulness of drinking tea, mindfulness of walking, mindfulness of breathing and looking up at the sky.

Check Out, Unplug,
and Take a Stroll in the Ultimate

—T.D.

Cultivating reverence for the simple wonders of life is, in our times, a powerful act of resistance. Choosing to step outside, to open our eyes, ears, and hearts to the presence of this beautiful planet takes courage and freedom. Society has conditioned us against it.

Thay's days in his hermitage in Plum Village were an artful balance of action and non-action. Whether he was working on a translation, doing research, preparing for a talk, or writing an article or letter, every few hours Thay would take a short, gentle stroll outside. In wind, rain, snow, or sunshine, he would stand up from his desk, fully aware of every step and breath, reach for his coat, hat, and scarf, and step outside to enjoy walking meditation in the garden, past the bamboos and pines and along the little creek running through. On a sunny day, Thay would lie in a hammock strung up between the trees and gaze at the poplar leaves fluttering in the wind. Sometimes an insight might come to him and, returning to his hermitage, he would pick up a calligraphy brush and capture the insight on paper.

There may be work to be done. When is there not work to be done? Thay teaches that, as a practitioner of meditation, it's up to us to assert our right to be free: to be a simple human enjoying being alive on a beautiful planet. Whenever Thay was invited to speak at a congress or parliament, he always insisted on leading an outdoor walking meditation at the end of the program. At Harvard, at Google, and at the World Bank headquarters in Washington, DC, it was the same: Thay wanted everyone to touch the peace and freedom of being aware of every breath and step and being fully present while walking through familiar streets, gardens, and plazas. The ultimate is not something far away. It can be available to us, in our deeply felt experience of the life around us, right where we live and work.

When I was working in London as a young journalist for BBC News, I had already begun studying with Thay in Plum Village. I asked a nun how I could keep my practice going when I got back to the city, and she told me I needed to create "islands of mindfulness" in my day. She recommended I get off the bus early and walk more of the distance to work. "You just need to choose a stretch to walk in mindfulness. It won't add many minutes to your day, and will keep your energy of mindfulness alive."

I chose a shortcut through a churchyard, and every day, crossing the street and stepping through the gate, I entered a realm of full awareness. I could hear the traffic, see the trees, listen to the birds, feel the pulse of the metropolis, and followed every single step and breath. Sometimes, if my thoughts were overtaking me, I'd pull up short, stop right there, take a deep breath, and recalibrate the pace. I have never felt so close to the soul of the city as I did in those few minutes of crossing the yard.

One of Thay's close friends and students is the remarkable Zen teacher Dr. Larry Ward. He writes about the healing power of being with Earth's wonders in his recent book *America's Racial Karma:*

An Invitation to Heal: "When I am in the natural world outside, I am moved by the experience of being non-judged and unharmed by the politics of my skin. I told a friend recently that I have never been disrespected or intentionally caused to suffer by a tree or a rock. I touch the wonders of life daily, and in doing so, I nourish my heart and mind with the flow of beauty, vastness, and gratitude as they rebalance wellness in my nervous system."

Cultivating a daily, regular practice of being outside "the four walls" of his house in nature, Larry explains that "a new and deeper vitality is now available to me to create myself and the world anew . . . To overcome injustice, we must not lose our centeredness, our spiritual resilience, and most importantly not our capacity to respond with wisdom, compassion, and action in creating a new world."

The Art of Non-Violence

The word for non-violence in Sanskrit is *ahiṃsā*. It means not-harming, not causing harm to life, to ourselves, and to others. The word "non-violence" may give the impression that you're not very active, that you're passive. But it's not true. To live peacefully with non-violence is an art, and we have to learn how to do it.

Non-violence is not a strategy, a skill, or a tactic to arrive at some kind of goal. It is the kind of action or response that springs from understanding and compassion. As long as you have understanding and compassion in your heart, everything you do will be non-violent. But, as soon as you become dogmatic about being non-violent, you're no longer non-violent. The spirit of non-violence should be intelligent. A police officer can carry a gun with non-violence because if they use their calm and compassion to solve difficult situations, they don't need to use the gun. They may look as though they're ready to use violence, but their heart and mind can be non-violent. It is possible to arrest, handcuff, and imprison a criminal with compassion.

Sometimes non-action is violence. If you allow others to kill and destroy, although you are not doing anything, you are also implicit in that violence. So, violence can be action or non-action.

Non-violent action is also long-term action. In the realm

of education, agriculture, and art, you can introduce non-violent thinking, non-violent action. Helping people remove discrimination is a fundamental action of non-violence because violence comes from discrimination, from hatred, fear, and anger. Discrimination itself is a kind of violence. When you discriminate, you don't give the other person a chance; you don't include them. And so, inclusiveness and tolerance are very important in the practice of non-violence. You respect the life and dignity of each person. Helping people transform discrimination, hatred, fear, and anger, before they become action, *is* non-violent action, and it's something you can begin doing right now. Don't wait to be confronted by a difficult situation to decide whether to act violently or non-violently.

Non-violence can never be absolute. We can only say that we should be as non-violent as we can. When we think of the military, we think that what the military does is only violent. But there are many ways of conducting an army, protecting a town, and stopping an invasion. There are more-violent ways and less-violent ways. You can always choose. Perhaps it's not possible to be 100 percent non-violent, but 80 percent non-violent is better than 10 percent non-violent. Don't ask for the absolute. You cannot be perfect. You do your best; that is what's needed. What is important is that you're determined to go in the direction of understanding and compassion. Non-violence is like a North Star. We only have to do our best, and that is good enough.

Violence and war do not always involve weapons. Every time you have a thought that is full of anger and misunderstanding, that is also war. War can manifest through our way of thinking, speaking, and acting. We may be living at war, fighting with ourselves and those around us, without even knowing it. There may be a few moments of cease-fire, but most are moments of war. Don't transform yourself into a battlefield. Suppressing or resisting your feelings can also be a kind of psychological violence. In Buddhist meditation we train to be there for our suffering, anger, hatred, or despair. Allow the energy of mindfulness to gently embrace and penetrate whatever feeling is there. You *allow* it to be, you embrace it, and you help it transform.

Even an economic system can be very violent. Although you don't see guns and bombs, it is still utterly violent because it's a kind of prison that prevents people from being included. The poor have to be poor forever, and the rich people rich forever because of the institutional violence in the system. We have to abolish that kind of economic system in order to include everyone and give everyone a chance for education, a job, a chance to develop their talent. That is applying non-violence in the realm of economics. When a business leader practices non-violence in their enterprise, not only will everyone around them profit from it but they will profit, too. It's not by having a large Gross Domestic Product that society can be happy; it's by producing compassion. You have the right to pursue economic growth, but not at the expense of life.

Non-Violent Resistance

As a young monk, there was a time when I was tempted to become a Marxist. I saw that, in the Buddhist community in Vietnam, people talked a lot about serving living beings, but they didn't have any practical methods to help the country, which was under foreign rule, and the people suffered deeply from poverty and injustice. I wanted to help create the kind of Buddhism that could reduce the social injustice and political oppression. I saw that the Marxists were really trying to do something, and they were ready to die for the sake of humanity. So, temptation at that time for me was not fame, not money, nor beautiful women; it was Marxism.

I did not become a communist because I was very lucky. I realized quickly that, being a member of the communist party, I would have to obey the orders of the party and may have to kill my countrymen who did not agree with the party, instead of being able to serve them. As a young person you are full of good intentions to serve your country, so you join a political party. You want to serve, not to harm, but your party may become like a machine, and one day you may be given the order to kill—or eliminate—other young people who don't belong to your party. And you have to betray your first intention to love and to serve. I was saved by the realization that violent revolution was not my path. I did not want to go in the direction of violence.

The principle of not harming, not killing, is very important. You try to help, to save others because you have compassion in your heart. Compassion is a powerful energy that allows us to do anything we can to help reduce the suffering around us.

In engaged action, you don't have to die to get your message across. You have to be alive in order to continue. We can get sent to prison, we can protest. But, even if our protest is very strong, we have to remember that protesting may not be able to remove the fear, the anger, and the craving in those we are protesting against. Real protest is to help them wake up and take up a new direction. To do that is real action. We can do it by setting an example. You create a community of peace and true solidarity. You consume in a way that protects the planet. You speak and listen in a way that transforms anger and division. You live simply and yet you are happy. This is a radical way of peace-making. You embody good health for yourself and for the world and live in such a way that proves a future can be possible.

Not Taking Sides

During the Vietnam War there was a lot of fear, anger, and fanaticism. The communists wanted to destroy the anticommunists, and the anticommunists wanted to destroy the communists. We imported foreign ideologies and weapons, and soon brothers were killing brothers. The two blocs were being supported by international armies, money, and weapons. Each side was convinced that their view was the best kind of view, and they were ready to die for their view. And yet there were many of us in Vietnam who did not want the war, and we tried to speak out. You were not allowed to voice your desire for peace because both warring parties wanted to fight to the end. And so, the peace movement had to be underground, and you risked your life if you spoke out for peace. Together with a group of young people, we distributed peace literature, and I wrote antiwar peace poetry, which was banned, and so we published it secretly. You risked arrest if you were found with a copy.

Speaking out for peace, we did not take sides. It was very difficult, very dangerous to take such a stand. When you take a side, at least you're protected by one side. But if you don't take sides, you're exposed to destruction by both, and so it's very difficult. And, in those conditions, we struggled for peace and engaged in social work in the spirit of non-violence and non-discrimination. It was very hard. It was in such a situation that our School of Youth for Social Service was misunderstood by both sides.

One night, armed men broke into the compound, kidnapped five of our social workers, and took them to the banks of the Saigon River. There, they asked many questions to confirm if they were social workers in our school, and when they said yes the men replied, "We are sorry. We've received orders to shoot you." And they were shot on the spot. The atmosphere of hatred was intense. We know the story because one of the students fell into the river when he was shot but survived.

We kill each other because we do not know who we really are. In order to kill someone, first of all you have to give them a label: the label of "enemy." Only if we see someone as our enemy can we shoot them without hesitation. But, as long as we still see they are a person, another human being, we can never pull the trigger. And so, behind violence and killing is the idea that the other person is evil, that there is no goodness left in them. Our view is clouded by hatred. We believe the other side to be a villain. And yet that villain is only a view, an idea. In Buddhism, the sword of insight is, first of all, to cut off the view, the label: in this case, that a person or a group of people is "evil." These labels are dangerous. They have to be cut off. Views can destroy human beings; they can destroy love.

Our enemy is not other people. Our enemy is hatred, violence, discrimination, and fear.

It was a very difficult, very painful time. The attackers had said "We are sorry" before they shot the social workers. And so, we know they did not want to kill them; they were forced to. The attackers were victims too. Perhaps they would have been shot themselves if they did not do it. And so, in the eulogy at the funeral for the social workers, the leaders of the School of Youth for Social Service said clearly that, despite the attack, they still did not see the killers as the enemy. There were no more attacks on the school after that. Maybe they were following us closely and heard what was said.

Many people misunderstood us, yet we still continued with our path because we had faith in our values. We had learned the truth—that the root of suffering and violence is intolerance, dogmatism, and attachment to views. In such a situation it is very important to not be attached to views, doctrines, or ideologies—including Buddhist ones. This is very radical. It is the lion's roar.

Bodhisattva of Respect

There is a bodhisattva whose name is Sadaparibhuta, the bodhisattva of constant respect, who never underestimates or disparages anyone. The action of that bodhisattva is to remove the complex of worthlessness and low self-esteem. This bodhisattva acts to bring the message of hope and confidence and remind all of us that we are a wonder of life. Sadaparibhuta can see the seed of awakening in every person. Even if you are disagreeable, Sadaparibhuta still smiles and says, "Well, even when you're shouting at me, even when you're angry, I still believe there's a Buddha in you." He is just trying to tell the truth. That is his vow: to go to everyone, rich, poor, intelligent, or less intelligent. And he always says the same thing. "This is what I really believe. I want to bring you that message that there is a Buddha in you. You are capable of understanding and loving."

All of us in our life as a human being will have to experience a moment of humiliation at one point or another. I have also gone through that. We may be victims of discrimination, victims of abuse, victims of injustice. But, with the insight that—no matter what—you still *have* your buddha-nature within, you are free. You are free from the feeling of being a victim. And you can be a bodhisattva equipped with enormous energy, with the power of changing your life, and even the power of changing the lives of those who have harmed you.

Does this mean that even those who commit atrocious crimes also have compassion? Is compassion innate? According to Buddhist psychology, we all have the seed of compassion but we all also have a seed of violence. We can envisage consciousness as having at least two layers. Below, there is "Store Consciousness" and above there is "Mind Consciousness." In our Store Consciousness there are many kinds of seeds—among them, the seeds of violence and of cruelty. And there is also the seed of compassion.

If you happen to be born in a kind of environment where people are compassionate and water the seed of compassion in you, then your seed of compassion will grow bigger and you will become a compassionate person. But, if you lived in another kind of environment, where nobody knows how to water the seed of compassion in you, then the seed of compassion in you will be very small. And if, for example, you watch a lot of violent films and live in a very violent environment and people water the seeds of anger and violence in you, then you will become a violent person. People see the violence in you and they can't see the compassion in you because your seed of compassion is very small. You cannot say the seed of compassion is not innate; it *is* there, but it has had no chance to be watered. That is why the practice of mindfulness consists of watering the seed of understanding and compassion every day, so we can restore balance between compassion and violence.

Compassion Protects You

When you witness a lot of violence, discrimination, hatred, and jealousy, it is your understanding and compassion that protect you. You are equipped with wisdom, the insight of interbeing, the wisdom of non-discrimination. And with that energy you can be a bodhisattva, helping others be more understanding and compassionate, including those who are causing harm. You are no longer *their* victim. You train yourself to see the seed of awakening in them, and you live in such a way to help them remove their discrimination, violence, and hatred. It is *they* who are the victims of their ignorance and discrimination, and *they* are the object of your work and practice.

Compassion can protect you better even than guns and bombs. With compassion in your heart, you will not react in fear or anger and you will attract much less danger to yourself. If you're angry, you make others afraid, and when they're afraid, they attack because they're afraid you will attack first. So, compassion protects both you and the other person. If you can produce compassion and prevent violence, that is a victory—a victory for both parties. It is a real victory.

When we are surrounded by turmoil and suffering, we have to practice taking refuge in our buddha-nature. Each of us has that capacity of being calm, being understanding, being compassionate. We should take refuge in that island of safety within to maintain our humanness, our peace, our hope. You become an island of peace, of compassion, and you may inspire other people to do the same. It's like a boat crossing over the ocean: if the boat encounters a storm and everyone on the boat panics, the boat will capsize. But, if one person in the boat can remain calm, they can inspire others to be calm, and there will be hope for the whole boat.

Who is that person who can stay calm in the situation of distress? In my tradition of Buddhism, the answer is *you*. You have to be that person. You will be the savior of all of us. This is a very strong practice, the practice of bodhisattva, taking refuge. And in a situation of war or injustice, if you don't practice like that, you cannot survive. You will lose yourself very easily. And if you lose yourself, we have no hope. So, we count on you.

Agents of Peace

—T.D.

What might a path of cultivating non-violence look like for you, in the way you interact with the living world, the way you speak and engage, the way you drink a cup of tea and consume? Reverence for life is a seed in our consciousness, and the stronger that seed is, the quicker it will be there in a difficult moment when we need it most. It takes insight, honesty, and courage to be able to say: *This tree is precious, this life is precious, this person in front of me—no matter what their views or values are—is also precious and is a child of the Earth just as I am.*

I find the story of Cheri Maples deeply inspiring. Cheri was one of Thay's senior students, a police officer who brought mindfulness to the police force and criminal justice system in Madison, Wisconsin. I remember meeting her under the shadows of the Rocky Mountains on a retreat in 2011—she was sharp, strong, and formidable, with bright eyes that were at once fierce and tenderly compassionate, as only a bodhisattva's can be. Cheri was a truth-teller and a fearless spirit. In her life and action she demonstrated that, with a strong personal spiritual practice, and a community to take refuge in, it is possible to realize far more than we ever thought possible.

Cheri's first retreat with Thay was transformative; she loved the meditation, she loved the spirit of community—and was determined to continue to practice when she went home. But the practice of non-violence and "not killing" seemed irreconcilable with her job; she carried a gun for a living. When she asked Thay about it, the response was "Who else would we want to have carry a gun but somebody who would do it mindfully?" Compassion can be gentle and compassion can be fierce. And, as Cheri learned from Thay, "Wisdom is knowing when to employ the gentle compassion of understanding and when to employ the fierce compassion of good boundaries." Cheri became a real "peace officer" and, over the course of her career, took her insights further, working to change and shift the culture around racial profiling, militarization, and police standards for using deadly force.

Cultivating Reverence for Life

We live in a violent society. We may experience violence on the streets, or at home, or as soon as we turn on our screens or the news. What we hear, watch, and read can touch off the seeds of fear, hatred, discrimination, and violence in us—often without us even realizing. And so, as a meditator, the challenge is to see whether we can be present and alert enough to notice when it's happening. Can we recognize the small actions in our everyday life that somehow contribute to war? Are we participating in, or privileged by, systems that are built on violence? What would we need to do or change in order to help create a just and regenerative culture that respects all life?

Respect starts with the basics, and the intention to actively cultivate a mind of non-violence in daily life. When Plum Village began offering live-streamed online retreats in 2020, we asked retreatants

to create a sacred corner in their home where they could follow the talks, meditations, and relaxations. With the pandemic, we couldn't welcome people to our meditation halls, bamboo groves, or old oak forests, but we could help them create a space that was sacred and inspiring, right in the heart of their daily life. That space is part of the architecture of our practice, watering seeds of peace and respect in our consciousness. Flowers are good, as are candles, perhaps incense, also elements from nature—maybe a rock or a beautiful autumn leaf. And it's lovely to have photos—of people who inspire us, our grandparents or ancestors, or places of the planet that hold a special meaning. When we invest in a space like this, it will be there when we need it most—when we need somewhere to sit and breathe and connect, to hope and dream, or simply to cry. Such a place can help us touch what is sacred and spiritual in our own life.

Here is the text of the mindfulness training on cultivating non-violence in daily life, the first of the Five Mindfulness Trainings. It is not a philosophy but quite literally a training: something we train toward. Having read the text, you may like to take a few moments to pause and reflect on how it lands in your heart. Does it resonate with you or does it trigger reactions? The text is designed to be an object of contemplation, to be reread from time to time, as a mirror to reflect on our everyday life and actions. May these phrases serve to challenge and strengthen your intentions, and inspire you to cultivate more reverence in your life.

The Mindfulness Training on Reverence for Life

Aware of the suffering caused by the destruction of life, I am committed to cultivating the insight of interbeing and compassion and learning ways to protect the lives of people, animals, plants, and minerals. I am determined not to kill, not to let others kill, and not to support any act of killing in the world, in my thinking, or in my way of life. Seeing that harmful actions arise from anger, fear, greed, and intolerance, which in turn come from dualistic and discriminative thinking, I will cultivate openness, non-discrimination, and non-attachment to views in order to transform violence, fanaticism, and dogmatism in myself and in the world.

DEEP SIMPLICITY:
YOU ARE ENOUGH

Reconsider Your Ideas of Happiness

To save our planet, we need to re-examine our ideas of happiness. Every one of us has a view, an idea, about what will make us happy. And, because of that *idea* of happiness, we may have sacrificed our time and destroyed our body and mind running after those things. But, once we realize that *we already have more than enough conditions to be happy*, we can be happy right here and right now.

Awakening is not something far away. Breathing mindfully, bringing our mind back to our body, we know we are alive, we are present, and life is there for us to live. That is already a kind of awakening. We don't have to struggle. We don't have to run into the future. We don't have to look for happiness in another time or place. We have to really be there *in* the present moment and live deeply in this moment where we can receive nourishment and healing. And, as soon as we touch that happiness, we no longer feel the need to fight or worry, and we have plenty of happiness to share with others. This is why collective awakening is so important.

We may wonder why corporate and political leaders are not doing more to save the planet. They know that the planet is in a dangerous situation that needs our urgent attention, and they may even want to do something to help. They are intelligent and well informed. So, it's not that they don't *want* to do something, but they find themselves in a situation where they *can't* do anything because they have their own difficulties.

They have pain and suffering, which they don't know how to deal with, and that's why they can't help us solve environmental problems. They are caught up in their own world. And they are caught in the view that it's money and power, first and foremost, that can bring happiness and help reduce suffering. But it's not true that wealth and power and economic growth means there is less suffering; many people have a lot of these things and still suffer. And so, we need to help them change their *idea* of what happiness is. And even that may not be enough; we need to give them a *taste* of real happiness. Only when they have tasted real happiness will they change their way of thinking and way of life and do business in a different way that helps protect the Earth.

To be happy is to be understood, to be loved, and to have the power to understand and love others. Someone without understanding and compassion is utterly cut off. Even if you

ZEN AND THE ART OF SAVING THE PLANET

have a lot of money, a lot of power, a lot of influence, without understanding and compassion, how can you be happy?

True happiness is grounded in freedom—not the freedom to destroy our own body and mind, nor the freedom to dominate and destroy nature, but the freedom to have time to enjoy life: freedom to have time to love; freedom from hatred, despair, jealousy, and infatuation; freedom from getting so carried away by our work and busy-ness that we no longer have time to enjoy life or take care of each other. Our quality of being depends on this kind of freedom.

The Present Moment Is a Whole World to Discover

—T.D.

Pursuing relentless economic growth will not guarantee our happiness, and may even threaten it. Happiness is not something we *attain* by accumulating wealth or status; it is something *available* to each of us, right in this moment, if only we can awaken to it. And yet, we have a tendency to not give the present moment much credit. We see a tree, and it's just a tree, doing not much. What's the big deal? The sky—very nice, but what's it got to do with anything? We've got places to go, things to do, problems to solve. There's a moon—we may think, "lovely"—and move on, get on with our life. How many moments and how many breaths did you give the moon when you saw it last? Can you tell me something about the first tree you see when you step out of wherever you live? What's its character? When does it sing its life song the brightest? Is it a song of blossoms, or buds, or burnished autumn leaves?

Thay once invited us to "draw open the curtains" on the present moment. The truth is that sometimes I find myself giving it just a small slice of my attention, perhaps 10 percent. I'm full of everything that's been happening (the past) and already anticipating everything I want to happen—or I am afraid of happening—in the future. We live in an overstimulated, crowded present. So as a practitioner of meditation, the challenge is to retrain ourselves to really allow slow-motion reality to soak into our consciousness. In the present moment, there's a multidimensionality we often neglect: there's touch, taste, scent, and embodied awareness.

As a meditator our task is to become fascinated with what the present moment *feels* like. I remember at first thinking that I need to access the present moment with my mind. But I've slowly understood that I can access it most directly with the senses: the smell of the woods, the edge of the breeze, the sound of the rain, the rumble of humanity through the asphalt. I've learned to cut through the noise and open up space to connect with a moment with my whole being. If the sun's looking spectacular, I remind myself to enjoy it with ten full in- and out-breaths, calming the body, savoring the senses. The same goes for the presence of a loved one or a beautiful tree. It takes strength to resist the running and just be there, open to whatever's going on inside and around us.

As a young novice I had a chance to be Thay's attendant on one of the days when he gave a talk. The task was simple but surprisingly challenging: just to be there, at Thay's side, at every moment, and assist him with whatever he needed—whether it was a coat, his glasses, a notebook, or, more often than you'd think, a cup of tea. And the challenge, of course, is to be aware of our own breaths and steps while we're doing it.

The first day was quite difficult. Thay is remarkably swift: every act so decisive and clean. He was already up reaching for his coat before I could help him, already at the door before I could open it. The training is to be both ahead and behind, and it was a lot to process; my mind kept getting in the way. I was mesmerized by his way of walking; he seemed to stop in every step and yet moved as effortlessly as a boat through water.

I remember one day accompanying him to his study after lunch with the whole community. I was glad to get to the doors before him, close them mindfully, put down the bag, and open up the hammock for him to rest. There was a peaceful moment of quiet as I slowly rocked the hammock and he gazed out the window. There was the sound of distant laughter, the song of the birds, the crackle of the stove, and the quiet ticking of the clock on the wall. "What time is it?" Thay asked, ever so gently. I wasn't quite sure. From where I was swinging the hammock, I couldn't see. "Erm . . . nearly two?" I offered, hesitantly. "I thought you were English!" he said, smiling, eyes bright as stars. I must have looked lost. Still he smiled. "Isn't it teatime? I thought in England it's always teatime!"

*t.d.

Open to Life

We can imagine there are many doors leading to happiness. Opening any of those doors, happiness will come to you in many different ways. But, if you are attached to one particular idea of happiness, it's as though you have closed all the doors except one. And, because that particular door does not open, happiness cannot come to you. So, don't close any door. Open all the doors. Don't just commit yourself to one idea of happiness. Remove the idea of happiness you have, and happiness may come right away. The fact is that many of us are attached to a number of things we think are crucial to our well-being—a job, a person, a material possession, an ambition—it could be anything. Even though we suffer a lot because of it, we don't have the courage to let it go. But the truth may be we continue to suffer precisely because of that. Each one of us has to look deeply and see this for ourselves. It needs great insight and courage to release our ideas of happiness. But, once we can do that, freedom and happiness can come very easily.

After I was able to return home to Vietnam in 2005, hundreds of young people asked permission to ordain as monastics in our Plum Village tradition. We were offered Prajna Temple in the central highlands, and there we began a new kind of mindfulness practice center training a new generation of Buddhist monastics. The temple grew so quickly, it was perceived as a threat to the authorities. In 2009, they tried to shut down the temple, and mobs were sent to intimidate and threaten the monastics. The four hundred monastics who lived there felt it was their home, their practice center, and they should do everything to hold on to it. Each one of them did their best to resist being forcibly disbanded. Prajna was for them a place where they could be themselves, speak the truth, and tell each other what was in their hearts. They wanted to cling on to that environment and community, no matter what. They resisted non-violently for more than a year and a half.

But, in the end, we realized that it was not the place that was important; it's the practice and the energy of togetherness, brotherhood, and sisterhood that's important. No matter where we go, we can take that with us. So the young monastics left and went into hiding and eventually found other places to continue to practice together. Today many of them are serving in our practice centers in France, Germany, and Thailand. In the process, we got something more valuable than a practice center: our faith in this path and in the community has grown stronger. We have now even found better conditions where it's easier to do what we want to do, to continue to train together in mindfulness and grow our community.

So, we should not hold on to an idea of happiness. If we're able to release our idea, there will be plenty of opportunities to realize what we want to realize. We should not be too sure of our ideas. We should be ready to let go of them. What we consider to be a misfortune may turn out to be a fortune later on. It depends on our way of handling the situation.

You Don't Need to Live
in a Cave to Be Zen

There are some who believe that, if they live in a remote place undisturbed by society, they'll have more time to practice. According to the Buddha, the best way to be alone and not be disturbed is to come home to yourself and become aware of what is happening in the present moment. During the Buddha's time there was a monk, known simply as Thera, who liked to be alone. He tried to do everything alone and was very proud of his practice of being alone.

Some of the monks went to report to the Buddha that there was such and such a monk practicing like that and claiming to be following the Buddha's teachings on solitude. So, one day, the Buddha summoned the monk Thera, invited him to sit down, and asked him if he enjoyed practicing alone. Thera said yes. And the Buddha said, "Tell me how you do it." And Thera replied, "I sit alone. I go to the village on alms round alone, I eat my lunch alone, I do walking meditation alone. I wash my clothes alone."

The Buddha then said, "That is one way of being alone. But I have a much better way of living alone." And then he continued, "Do not pursue the past. Do not run into the future. The past no longer is. The future is not yet there. One who dwells in mindfulness night and day is one who knows the better way to live alone."

To live alone (*ekavihārī*) means to not have anybody with you. That "body" may be the past, the future, or your projects. It may be the object of your seeking or craving or the idea of happiness you have. To live alone means to be completely satisfied with the here and now, to have a deep sense of fulfillment in the present moment. You don't need to go up into the mountains, or to a cave, to be alone. You could be on the mountain or in the cave and still be longing for something else, still searching, or still regretting—and then you're not alone. But with mindfulness, you can sit in the heart of the marketplace and still be alone and have peace and freedom. It doesn't take years in a cave.

Set Yourself Free

Freedom is a practice. It's not something we earn after ten years. As soon as we cut through our regrets and anxiety and get in touch with the present, we get freedom right away. All of us are warriors, and mindfulness is the sharp sword that sets us free.

Everything you are looking for, everything you want to experience, must all happen in the present moment. This is a very important point. The past is no longer there; the future is only a vague notion. If we grasp onto the future, we may lose the present moment. And, if we lose the present moment, we lose everything—our happiness, freedom, peace, and joy. And so, all our aspirations, all our dreams, all our projects have to be brought into the present moment and centered in the present moment. Only the present moment is real.

You simply breathe in and realize you have a body. You smile to your body; you enjoy having a body to sit or walk on the Earth and enjoy the Earth. And with that energy of mindfulness you also take care of any ill-ease, restlessness, or suffering you find there. This is very concrete. It's not a philosophy or an idea but a real way of practice to help you suffer less and enjoy life more—starting with your breath, your body, the Earth.

You realize your body is wonderful and contains all your ancestors and future generations. You feel the privilege of being alive. Life is not only suffering; it is also full of wonders.

Allow the artist in you to recognize, admire, and yearn for what is good and beautiful. The artist in you has never died. Every morning that you can enjoy the sunrise, that is the artist in action. The warrior goes hand in hand with the meditator and the artist. We also have to give the warrior in us a chance. Our warrior's weapon is the sword of wisdom that sets us free.

Life is wonderful, mysterious. There is so much to discover and be curious about. *Allow* yourself to be free to enjoy your time on Earth. Helping the world, liberating people from their suffering, is something we can do. But, first of all, we have to help ourselves and set ourselves free.

Who Is the Boss?

In our times, many of us live on autopilot; we live like machines. Living with mindfulness is different. When you drive, you know you are driving. You are the boss. The car isn't driving you; you are driving the car. And, when you breathe, it's not just because the body needs to breathe that you breathe. You are breathing in and you enjoy breathing in. Freedom is cultivated by small things like that. When you walk, it's not just to get somewhere. With every step you enjoy being alive. You are truly yourself. And that is freedom. With mindfulness, freedom becomes possible. And, the more freedom we have, the happier a person we become.

If in the twenty-four hours of your day you have five minutes of peace, ease, and freedom, that's not too bad. See how generous we are! When you have just five minutes without letting craving, projects, or anxiety carry you away, you become a free person with nothing to do and nowhere to go.

And yet, many of us get swept away by events and situations around us and by whatever we see and hear. We *lose* ourselves. That is why we must also cultivate freedom from the crowd. When you have real freedom, even if the crowd is shouting or going in one direction, you can still be yourself. You're not swept away by the emotions of the majority. You need to be very strong to have that kind of freedom.

The Buddha was a monk who had great freedom. Even

when everyone else thought differently, he could still see that his way of thinking was the truth. For example, when the Buddha talked about no-self, many people objected because that insight went against the whole way of thinking in India at that time. Yet the Buddha had the courage to uphold his insight. He had freedom inside, and he was patient. And, finally, he could make those insights available for many other people. True freedom brings love, patience, and many other wonderful qualities.

This Is It

When I was a novice, I accompanied my teacher on a visit to Hải Đức Temple in Huế, Vietnam. There, I saw a Zen master sitting on his wooden platform. The image struck me. He was not doing sitting meditation. He was not in the meditation hall. He was simply sitting in front of a low table, very beautifully, very straight. And I was very impressed. He looked so peaceful, natural, relaxed. And in my heart as a novice, there came a vow, a deep longing, to sit like that. How could I sit like that? I would not need to do anything. I would not need to say anything. I would just need to sit.

In the Zen tradition, meditation is considered a kind of food. You can nourish and heal yourself with meditation. In Zen literature, you can find the phrase "the joy and happiness of meditation as daily food." In Chinese it is written: 禪悅為食.

Every session of sitting meditation should give us nourishment, healing, and freedom.

Practicing sitting meditation is like sitting on the grass in a spring breeze. In Plum Village, we sit so we can be truly present. We sit to be in touch with all the wonders of the cosmos, of life, in the present moment. There is no other aim. We just sit, and we don't need to do anything. We don't even need to get enlightened. We just sit to be happy. We sit to have peace and joy. Sitting is not hard labor. In the Soto Zen tradition, they speak of "sitting for the sake of sitting," or "simply sitting." We don't sit to *do* something. You only need to sit.

We need to organize our daily life so we have more opportunities to be, to learn being peace, being joy, being loving, being compassionate. We need very concrete ways to do this. How can we stop being victims of overscheduling? Our society is so caught in our daily concerns and anxiety we don't have time to live our life or to love. We don't have time to live deeply and touch the true nature of what is there, to understand what life is. We are too busy to have the time to breathe, to sit, to rest.

Why do we need to be so busy? You have to accept the fact that it is possible to lead a simpler life—a life that gives you more freedom. You should live a life that can allow you some time to sit and do nothing. And, when you sit quietly, you can begin to see many things. You have time to take care of your body, to take care of your feelings and emotions. You taste the joy of being free.

The Courage to Sit

—T.D.

The problem about coming back and landing in the present moment is what we find when we get there. Do you think that's why we avoid it so much? As soon as we come back to our body and close our eyes, we discover we're full of everything we've been experiencing and all the images, sounds, and feelings that come with it. If the world feels broken already, why would we want to feel that more? We don't want to encounter it; we want to *resist*.

It may seem that there is a paradox here: on the one hand, the Zen masters tell us to breathe and accept the situation; on the other, they say we must seek to change it. The way out is to do both. How can we hope to change something if we have not yet understood how it came to be? How can we listen and understand what's going on outside if we can't hear and understand what's going on inside? Twenty or thirty minutes of sitting can be twenty or thirty minutes of taking care of the world, in the way its suffering is reflected in our own body and feelings. It takes courage.

In Plum Village, we practice sitting meditation for thirty minutes twice a day. Already walking along the path to the meditation hall our sitting meditation begins: following our breathing, arriving in every step. We open the door, remove our shoes, and are fully present, feeling every step as we approach our cushion to sit down. It's important to find a stable, comfortable posture: we sit with our body, not our mind. Many of us start with a body scan, relaxing every muscle with the gentle energy of mindfulness. There's already the act of arriving and listening to how the day is playing out in our brow, jaw, shoulders, or chest.

We train to encounter ourselves with gentleness and without judging or reacting. We don't sit to be a buddha, to be someone else, someone better, or someone different. We just sit to be ourselves, sitting. Creating a window of fifteen minutes of freedom to be ourselves every day is already something. There's an art to sitting. It's not about adding up the minutes, holding a certain posture, or escaping somewhere other than right where we are. There's an ease, a naturalness, and an aimlessness to just being there, alive, fascinated by the miracle of breathing and sensing the world.

Recorded guided meditations can be helpful to keep the thread of our concentration if we're distracted, but a direct, silent experience of our own present moment can be even better. We listen deeply to the imprint of the world in our body and feelings; we dissolve the restlessness, soothe the anxiety. If necessary, we cry the tears. We don't meditate only to touch peace; we meditate to recognize, embrace, and transform everything we discover is blocking the way between us and peace. Thay always says, "You're allowed to cry. Just don't forget to breathe"—we embrace our tears with the energy of mindfulness. In sitting, we need the compassion and creativity of an artist, the stillness of a meditator, and the discipline of a warrior. We need a strategy. Where will you sit? When will you sit? Everything you do leading up to the sitting is already the sitting.

The Power of Simple Living

As a young novice in my temple, we had no running water, no hot water, no electricity and yet we still lived very happily. Even washing the dishes for a hundred monks was very joyful because we did it together, with the joy of brotherhood. We would walk up into the hills to rake pine needles for the fire to heat the water. We had no soap but used the husks of coconut and ash to clean the pots. Happiness doesn't depend on external conditions alone; it depends on our way of looking at and seeing things. We can live more simply if we know how to cherish the conditions we have.

Mahatma Gandhi said that we must be the change we want to see in the world. If we know how to live a more simple, relaxed, and happy life, our planet will have a future, and all species on our planet will have a future. It's a dream we can realize right now, today. Gandhi dressed simply, walked on foot, and ate frugally. The simplicity of his life is witness not only to his freedom from material things but also to his spiritual strength. What made Gandhi's struggle a great success was not a doctrine—not even the doctrine of non-violence—but Gandhi's own way of being.

People everywhere are trying to apply the principle of non-violence but struggle to reproduce Gandhi's vitality. Lacking Gandhi's spiritual strength, it is hard to produce his level of compassion and sacrifice. But, so long as we continue to allow ourselves to be dragged along by the machinery of consumerism, it will be difficult to build our spiritual strength. This is why the point of departure for a new civilization must be a determination to not be "colonized" by material goods—to rebel against materialism in favor of our need to be a human *being*.

Living more simply can give us tremendous courage to speak the truth. When we speak out, we do so out of love and compassion. You have the courage to speak out because you're not afraid of losing anything. You are free. You know that understanding and love are the foundation of happiness, not material possessions, status, or position. But, if you're afraid of losing these things, you won't have the courage to speak out.

Bodhisattva Samantabhadra

Bodhisattva Samantabhadra is "the bodhisattva of great action." There are many kinds of action we can take in the spirit of Samantabhadra to help relieve suffering in the world, including the practice of generosity, giving. Our daily life should be an offering. You don't need a lot of money to make an offering; your own peace and happiness are already a big gift for others. You may feel generous, but you must also find specific ways of developing your generosity. Time is more than money. Time is life. Time is for being deeply present with the other person. Time is for bringing joy and happiness to others.

Your presence—your way of being—is what you offer every minute, every hour of the day. Samantabhadra is not an abstract figure. Samantabhadra is flesh and bone all around us. Samantabhadra is there in each one of you who is doing something to bring relief to people around the planet. Even in my own community, I see many bodhisattvas working tirelessly to help others, and I feel very grateful to them. Some are young, some are less young. They are all arms of the bodhisattva of great action. When we help, we don't feel forced to do it; we're glad to do it. Our practice is to live our daily life in such a way that every act becomes an act of love. We serve all beings with our understanding, compassion, and action—and we can have happiness right while we are acting.

What Should I Do with My Life?

How can you know you're living a life that can contribute the most to humanity? How can we choose the kind of job where we can be more at peace with ourselves and more helpful to the world?

Deciding what to do is a question of deciding how we want to be. Doing is a way of being. What is essential is that, while you do it, you enjoy it and you are fully offering your presence to the world and to yourself. Anything is good. It depends on how you do it, not what you do. There are many kinds of jobs that can help you express your joy and your compassion for humankind and for all species. You may earn less, live in a smaller house, and have a more-humble car, but you'll be happier. You can laugh and you can love—everything you do is an expression of love. If you can live like that, your life is a happy life; simple living is possible. The most helpful thing for you to do is to be a happy human being.

Many of us are being overtaken by work and we are losing our life; we do not have time to live our life deeply. That is our civilization. Burnout is a reality. It's not worth it. Life is a gift and we need to make the most of that gift. We need time to live. We need to avoid a kind of dualistic thinking that "work" is one thing and "life" is another thing.

When you chop wood, when you carry water, when you cook breakfast—all these things are "work," but joy and happiness are possible during the time you do the work. When you have a meeting with a client it is the same. The meeting is not only to get the deal but can be transformed into a pleasant encounter between two living beings. The element of mindfulness and loving kindness can make that meeting a joyful, meaningful, and happy moment. It is a question of quality not quantity, of *how* you live, not how much you do or how efficient you are.

Whenever I do calligraphy, I practice to see life and work as one. Beginning each session, I always prepare tea, and then I mix in some tea with the ink. Tea and Zen have gone together for many thousands of years. And, as I draw a half circle with the brush, I breathe in. As I draw the other half circle I breathe out. There is breathing in the circle; there is mindfulness. From time to time I invite my teacher or my father to draw the circle with me. I touch the truth of no-self because my teacher is in me, my father is also in me, and the meditation, the work, joy, and life become one. There is no distinction. The art is how to make your work pleasant and enjoyable— and that is possible with mindfulness, and with training.

How Can I Make Difficult Decisions?

In our daily life we often find ourselves in a situation where a decision has to be made, and we make the decision too quickly, or when our mind is anxious or agitated. That decision does not come from a clear mind. We should avoid making decisions when our mind is not free. Even if people pressure us, we should refuse because a wrong decision can make ourselves and others suffer for a long time.

So, don't make the decision. Breathe in first. Breathe in and concentrate entirely on your breathing, let go of what is preoccupying you, release the past and the future, and you become freer. Within five or seven minutes you may already have enough freedom to make the decision. Your breathing nourishes what freedom you have, and there will be a variety of decisions available to you. Your decision will be better, more beneficial, more compassionate than a decision taken under the influence of fear, regret, anxiety, or sorrow. You can get a *lot* of freedom when you breathe mindfully. And, if you want that freedom to continue, you can extend your mindful breathing as long as you want. The practice is simple but very effective.

To practice mindfulness doesn't mean you're forbidden from making plans for the future, and it doesn't mean you're prevented from learning from the past. The idea is not to get lost in fear or uncertainty about the future but to be *grounded* in the present moment and bring the future into the present moment and have a deep look. That is planning for the future. You're not lost in the future; you are planning for the future right in the present moment.

How to Fail

Are you afraid of failure? How can you handle that fear? You may doubt whether you can realize what you want to realize. But what is success and what is failure? Every one of us is motivated by the desire to succeed. Some of us fail but, based on that failure, we do better and can get a kind of success that is truly success. And there are others who succeed but become victims of their success. That is not the kind of success we want.

Suppose you practice mindful breathing and you succeed. Mindful breathing brings the mind home to your body and helps you establish yourself in the here and now. It brings joy and peace. That kind of success can never harm you. It's a question of your *way* of doing things. If you want to succeed, and you do whatever you can to arrive at that success, even using means that are not good, you may "succeed," but you ruin yourself as a human being.

Right action is the kind of action that goes in the direction of understanding and compassion and truth. It is the kind of action without discrimination based on the insight of inter-being. Right action is motivated by compassion. And, if everything you do to arrive at your success is right action, then you have nothing to fear because that right action is generating the energy of goodness, of compassion, of peace. And that can protect you all your life.

Taking care of what you think, say, and do, there is no more fear of failure. With the energy of mindfulness, concentration, and insight, every moment of your life becomes a success already. You don't need to wait a year to succeed. Every step is a success. Every breath is a success. Because with every step you can generate joy, peace, and happiness. And, if the means are good, the end will be good.

The Student and the Hermit

The world needs light. We need people who can bring into this world the light of freedom, the light of understanding, the light of love. Buddha Dipankara is someone with the capacity to "light the lamp of wisdom" and shine light on the world.

Legend has it that, in a previous lifetime, the Buddha Shakyamuni was once a student. He wasn't yet a buddha; he was a bodhisattva. And he was a student, following higher studies, who dreamed of becoming a statesman. The young men of his day all had the same dream: to pass the exams and be chosen by the emperor to become a statesman. And so, everyone's parents and friends did everything they could to help students to pass the exams.

Each province organized competitions and, of the many thousands who entered, only a hundred would be chosen and sent to the capital for further training and selection. In the capital, the emperor himself set the questions for the dissertation in the Imperial Competition. He wanted to test whether the candidates understood the situation in the country and to see what ideas they had for helping the people and society develop and be happier.

The young man entered the competition but he wasn't chosen. He felt deep despair. He had studied hard to realize his dream of serving his people and his land, earning a good salary, and having a family. Disheartened, he began the long journey home on foot across mountains, forests, and fields. One afternoon, coming to a hill, utterly exhausted, he felt he couldn't go on. Just then he came across a hermit, a monk living very simply at the foot of the hill. He stopped and noticed that the hermit was cooking something in his little pot. He was so hungry and tired, he asked for something to eat. The hermit said, "Rest a little bit and, when the soup is ready, I will give you a bowl. Here, you can use the roots of this tree as your pillow." The young man lay down to rest and soon fell into a deep sleep.

He dreamed a very strange dream. In the dream, he saw that he had been one of the top one hundred chosen in the triennial competition and had been sent to the capital to enter the Imperial Competition. He did his very best to answer the questions asked by the emperor, using all the knowledge he had acquired through reading many, many books. He was chosen by the emperor and, since he was considered the most brilliant of all the young people who had presented themselves at the competition, the emperor offered him the hand of the princess. The princess was very beautiful, and you cannot imagine how happy he was. He was full of hope and full of energy. He was given a very important post in the cabinet as Minister of Defense.

But theirs was a small country adjacent to a much stronger country. As Minister of Defense he was responsible for defending their borders, and he encountered many difficulties and challenges, including jealousy, despair, and anger. His relationship with the princess was not easy either, and they argued nearly every day. Their two children were also very difficult to bring up. There was a lot of unhappiness. There were difficulties in his married life and in his political life.

One day, he got the news that the neighboring country had amassed large troops to invade. He had to summon his forces and dispatch them to the frontier to resist the invasion. Already struggling in both his public and private life, he didn't have enough peace and clarity in his heart, and so, when he organized the counterattack, he made many mistakes. The enemy prevailed and seized a lot of territory. News of the defeat reached the emperor and, furious, he gave orders to behead the Minister of Defense.

In the dream the young man saw himself surrounded by soldiers and taken to the execution block. Just as he was about to be beheaded, he heard something like the song of a bird, and he woke up. Disoriented, he looked to the left and right. He saw that he was at the foot of a hill, and near him was the hermit.

The hermit looked at him with a beautiful smile, and said, "Did you have a good rest? The millet soup is ready. Come and sit here and let me give you a bowl of this good soup." The young man got up and was barely hungry. He had seen so much in his dream it seemed a whole lifetime had passed. If you do not know how to live each moment deeply, life can pass you by like a dream, very quickly, perhaps even quicker than the time it takes to cook a pot of millet.

The hermit was there, calm and serene, stirring the soup with a pair of chopsticks. Looking deeply at the hermit you could see that peace was alive in him. He was really alive; he was really happy. With peace, with solidity and freedom, life is something wonderful, and happiness is possible. The young man sat close to the hermit and asked him many questions. Since he was intelligent, he began to discover that peace in the heart and freedom in the heart are essential to a happy life, and he gave up his ambitions to be a statesman. He formed a new dream. He wanted to learn to live like the hermit so he could transform his suffering and restore peace and freedom in his heart. He decided to become a disciple of the hermit.

This hermit was the Buddha Dipankara, the one who lights the lamp. And, afterward, the young man, when he had practiced through many lifetimes, became a buddha with the name Shakyamuni. If you are still a student, you may like to think about this story. Look deeply into your ambitions and plans, to see if it's worth spending all your life and your energy chasing after those things you crave.

To be in touch with someone, a friend, who knows the path of freedom, who knows how to generate solidity and compassion is something very necessary. There is a very important word in Buddhism: *kalyāṇamitra*. It means a friend who is wise, a friend who has light—a true spiritual friend. This friend might already be very close to you but you haven't been able to recognize them. We all need to find that friend who can support us and give us light so we don't lose our way in the dark. And the moment you find that spiritual friend is a wonderful moment.

Learning the Art of True Happiness

—T.D.

The great Zen masters tell us that, if only we take the time to be still and listen, we will know the path we need to take. No one else can give us that insight—it's something we have to get for ourselves, with our own authentic practice. Insights can't be transmitted in words, or even in a book. The words can point the way, yes, but they can't walk the path for us.

As a young journalist, I practiced mindfulness in the newsroom, on the top floor of a six-story block in central London. I trained myself to listen to my colleagues when they were yelling, to breathe before answering the phone, and to visualize mountain springs and waterfalls every time I went to get a cup of water at the cooler. When the going got tough—when a live guest dropped out at the last minute or a tape got lost—I trained myself to remember where the stars were: not only above my head, beyond the city's gray skies, but also to the left, the right, and under my desk, far, far away past the other side of the planet. In moments of extreme pressure, I discovered I could do short ten-minute relaxations in the bathroom cubicles. And I found it was possible to be present with every step as I sprinted down the studio hallway at the last minute to deliver cue cards seconds before going live on air.

Mindfulness helped me live my newsroom minutes deeply and helped me see my situation clearly. I remember one day following my breathing and my steps as I crossed the carpeted floor to make a coffee, my mind buzzing with the stories from the two radio shows and six newspapers I'd had to take in that morning. As I took a mindful step into the kitchenette, the question suddenly caught my breath short: Is *this* how I want to live my hours and days and months? I realized I was a tiny cog in a toxic machine, spread across thousands of open-plan studios, offices, and kitchenettes. Is this really what I want to do with my one, precious life? Is this where I want to invest my energy? The question planted itself right in the heart of my life, and became a koan. And, one day, the answer presented itself, clear as day.

Today's generation of outspoken young climate activists are a very special kind of bodhisattva—stark truth-telling heralds and mirrors on our world. Their extraordinary impact shows that we don't need an economics degree to speak the truth. The truth is there, right before our eyes, if only we allow ourselves to see it, hear it, and speak it. We need many more truth-tellers in our world, and we need many different kinds of bodhisattvas. What kind of bodhisattva will you be? How will you choose to spend your time and energy? As a species, the choices we're making about our time, energy, and livelihoods are at the root of our planetary crisis.

Here is the next of the Five Mindfulness Trainings, for us to take as a challenging object of contemplation to guide and accompany us toward a deeper simplicity and fulfillment. After reading it, you may like to take a moment to pause and reflect on what light it sheds on your life right now.

The Mindfulness Training
on True Happiness

Aware of the suffering caused by exploitation, social injustice, stealing, and oppression, I am committed to practicing generosity in my thinking, speaking, and acting. I am determined not to steal and not to possess anything that should belong to others; and I will share my time, energy, and material resources with those who are in need. I will practice looking deeply to see that the happiness and suffering of others are not separate from my own happiness and suffering, that true happiness is not possible without understanding and compassion, and that running after wealth, fame, power, and sensual pleasures can bring much suffering and despair. I am aware that happiness depends on my mental attitude and not on external conditions, and that I can live happily in the present moment simply by remembering that I already have more than enough conditions to be happy. I am committed to practicing Right Livelihood so that I can help reduce the suffering of living beings on Earth and stop contributing to climate change.

RIGHT FUEL: GUARD YOUR MIND, NURTURE YOUR ASPIRATION

What Are You Feeding?

In order to be able to really protect the environment, you have to be able to take care of yourself. The base of our operations is our daily life and our body, feelings, perceptions, mental formations, and consciousness. The well-being of the planet depends on the well-being of your own body and mind, just as the well-being of your mind and body depends on the well-being of the planet. And so protecting the planet has to do with our way of consuming. If you cannot deal with the problem of pollution and loss of balance in yourself, how can you deal with the problem of pollution and loss of balance in nature? The teaching of interbeing here is very important.

We suffer because we have been consuming the wrong kind of food. We are ruining our planet with our way of consuming, and our children risk suffering deeply. So, the way to save our planet is to practice mindful consumption. Otherwise, humankind will continue to ruin our planet and create a lot of suffering, not only for other human beings but also for other species on Earth.

The Buddha said, "Nothing can survive without food." And he also said, "When we suffer, we blame the outside, and we blame other people we think make us suffer. But, looking deeply, we find out that our enemy number one is ourselves." *We* are the ones who make ourselves suffer the most, by the way we consume, the way we eat, the way we drink, the way we organize our life, the way we behave—even by our way of chasing after our ideas of happiness. We are the ones who create our own suffering; we are our own worst enemy. That's what the Buddha said. In many ways we are *responsible* for our own suffering. We may have thought that something is good for us, and yet it makes us suffer a lot.

It is possible to end our ill-being. That's good news! Depression can be ended. Your fear, anger, and hatred can be ended. We can train ourselves to see our ill-being in terms of nutriment. As soon as we can recognize whatever is the source of nutriment for our ill-being, we just cut it off and our suffering will cease; it has no more fuel and it will die.

In Buddhism, we speak of four kinds of nutriments: edible foods (what we eat and drink), sense impressions (everything we consume through our senses in terms of images, sounds, music, movies, websites, and so on), volition (what we consume in terms of our deepest intention), and consciousness (what we consume in the collective energy around us). All these sources of nutriment can be healthy or toxic.

Deepest Desire

The first of the Four Nutriments to contemplate is "volition." What do we want to do with our life? We have to sit down and look deeply to find out. Is your deepest desire to run after fame, power, success, wealth, and sensual stimulation, or is it something else? For a terrorist, their deepest desire is to punish and to kill. For an ecologist, their deepest desire is to protect the environment.

We all have desire, and our desire can be healthy or unhealthy. It can make us suffer or it can make us happy. Is our deepest desire healthy or not? If our deepest desire is to suffer less and be happier; if our deepest desire is to come back to ourselves, to create joy and happiness, and nourish ourselves, and help others do the same; if our deepest desire is to learn how to embrace and transform our suffering, so we can suffer less and help others do the same—then that is good. That is a good aspiration, that is bodhicitta, the best kind of volition.

We know there is suffering in ourselves and in the world. We want to *do* something, to *be* something, to help reduce the amount of suffering that is there. But we may feel helpless because the amount of suffering is so overwhelming. And, alone, it doesn't seem we can do much. We can't bear to live anymore, even though we may still be very young.

When the Buddha was young, he had the same feeling. He saw the suffering, and he saw that, even if you are king, there's not much you can do to alleviate the suffering. And so, he chose not to become king, turned his back on the royal court, and tried to find another way. And what motivated him to become a monk, to practice, was a desire to help people to suffer less. Once we can transform the suffering in ourselves, we'll be able to help transform the suffering in the world. It's very simple, very clear. And that is what the Buddha did.

We practice mindfulness—we may even become monastics—not to avoid suffering nor to avoid society but to get the strength we need to cope and to help. And we learn that although alone we cannot do much, with a community we can do something. That is why the first thing the Buddha did after enlightenment was to look around for the elements to build a *sangha*, and he was an excellent sangha builder. Over the past decades I have also learned to build sangha. When I met Martin Luther King Jr., he used the words "beloved community" to describe the sangha, our spiritual community.

When I was a young monk, the suffering in Vietnam was

overwhelming. Millions of people died. And how can you help? You are overwhelmed by the suffering. And yet you want to do something to help end the war. My friends and colleagues—including Sister Chan Khong, who was still a young student at the time—did everything they could to help relieve the suffering of those who were poor and oppressed. But it wasn't enough; the war continued to destroy so much. And so, they began to engage in peace activities, and Sister Chan Khong even got arrested. I also suffered as a result of my efforts to bring peace: forty years of exile because I dared to do something to help bring the war to an end. Forty years of exile! But we had to do *something*; otherwise, we could never have survived, either physically or mentally. We go *insane* if we can't do something.

For all of us today, it is the same. Our planet is in danger. There is so much violence, so much suffering going on in the world, you go crazy. You want to *do* something, first of all just to survive and then to help reduce the suffering. You *aspire* to do something. You *desire* something. And you need that kind of desire in order to have enough energy to sustain yourself. Your deepest desire is not just to have money, social recognition, influence, or success. What you really want is something more.

Maybe you want to change the direction of civilization. Maybe you want to help people take care of themselves and their suffering so they can heal and transform, live deeply with joy and happiness, and help the Earth restore her beauty. That is a good desire, a good source of nutriment. That is bodhi-citta, the mind of love. And, if you are a politician, an activist, or a corporate leader and you have that kind of good intention, good volition, you can reverse the direction our civilization is going in.

There are desires that can destroy you, your body, and your mind. But there are also desires that can give you a lot of strength: an aspiration, a vow. And, as a young person, you need that kind of food. Of course, we have our own suffering, but as soon as we have a strong aspiration, we are ready to do something. Seeing the suffering in the world, right away we feel that the suffering in us is not the most important thing, and we suffer less right away. And that is why this kind of nu-triment, this kind of food is very important.

And, when you have this kind of desire in you, your eyes shine brighter and your smile is more beautiful and your steps become more firm. And that strong desire is the kind of nu-triment, the kind of food, that you need. And, when we come together as a community with a *collective* aspiration, we have the energy we need to realize what we want to realize. And we take refuge in a community, not for our own sake but for the sake of everyone because, without community, we cannot go far.

You know what to do. You have to master the techniques of mindfulness, concentration, and insight, because the young generation needs you. Your time is for that. Every moment is an opportunity for us to train, to transform ourselves, to prepare ourselves to serve the world. We have no time to waste on things that are not really important. Our path is very clear. We have something to do. And we know that, if we can do it, we can help reduce the suffering of the world. Even though I am advanced in age, I have been able to keep my aspiration alive. I'm still young in spirit and I want to transmit that energy to my students. Do not grow old; stay young. Get the appropriate kind of food. And build community together.

Bodhisattva Ksitigarbha

Any bodhisattva, any great being, always has a tremendous source of energy in them. If you don't yet have an aspiration, you need to find it. We should sit down with our partner, with our friends, and inquire about each other's deepest dreams. And if you share the same kind of aspiration, your relationship will get stronger. We are here, alive, and we all want to do something with our life. We want our life to be useful, meaningful.

There is a bodhisattva whose name is Ksitigarbha. The vow of Ksitigarbha is to go to the places where there is a lot of suffering in order to serve and to help. There are hells a little bit everywhere on the planet. Sometimes hell may be in our own family, our own community, or our own nation. We are hating each other, we are killing each other, we are using bombs and guns to kill each other. We belong to the same family, the same nation, the same community, the same culture, and yet we make a hell for each other. Ksitigarbha is ready to go to these areas and help. It's very difficult. You have to be fresh, you have to be patient and persevering, you have to be armed with non-fear. And you come not to blame but to help stop the fear, the anger, and the violence. You come to offer understanding and compassion and to help others generate understanding and compassion in their hearts.

There are many doctors, nurses, and social workers who are living Ksitigarbhas. They volunteer to be present in painful spots on our Earth to help. So, Bodhisattva Ksitigarbha is very real—they are not just an icon. Many young people are serving as Bodhisattva Ksitigarbha all over the planet. They are not afraid of suffering because they know that they can bring relief. They are protected by a strong energy of compassion and aspiration.

We should all support those who make a strong vow to help others so they will not lose their aspiration or burn out. We send them our energy of encouragement. So, after six months or one year working in a very difficult situation, they come home and need nourishment and healing. We are there to take care of them and try our best to help them heal so they can go out again to help a second time, a third time. The Ksitigarbhas in the world need help. They need a community, a sangha, so they can continue for a long time.

In difficult situations, you have to know how to cultivate your inner artist, meditator, and warrior—the determination and fearless endurance of Ksitigarbha—so you can be balanced and steady for your people. Many decades ago, there were times when it looked like the war in Vietnam would never end. Despair was prevalent, especially among young people. I remember they would come and ask, "Thay, will there ever be a day when the war is over?" It didn't look like it was ever going to end; it dragged on forever and ever. It was very difficult for me to give an answer, but, after breathing in and out I said, "My friends, we know everything is impermanent, and so is war."

The answer to that question is not so important. What's important is, in every situation, to find a way to cultivate compassion, calm, and clarity. If these can be kept alive, then there is hope. The worst enemy is despair. We must keep hoping. Our practice of calming and looking deeply nourishes our hope. And, with that calmness, that looking deeply, and that openness, we can grow a movement of awakening, in size and in quality. So many people are ready to be something and to do something for the causes of peace, of social justice, of protecting the planet. We should not feel alone. The temptation to despair and use violence is always there. But, if the meditator and the artist are alive in us, then the warrior will know exactly in which direction to go.

Is It Zen to Dream?

You might say Thay Nhat Hanh has always encouraged us to live in the present moment and now he's telling us to dream about the future! It might be a beautiful dream, but it's still just a dream.

What is our greatest hope and aspiration if not a dream? In Buddhism the energy of bodhicitta, the mind of love, is not just a dream. Bodhicitta is a reality, a living energy giving us faith and hope. Each moment of our daily lives, our dreams can slowly become reality. Over the past few decades, in my own life, I can tell you there's not a single moment when I haven't witnessed my dreams becoming real. Our dreams *can* become real. Indeed, they are *already* becoming real. Dreams may never be realized 100 percent, but each day they can slowly become more and more real, something so real we can touch them right in the present moment.

In Mahayana Buddhism cultivating a deep aspiration is a very important practice. In order to become a true bodhi-sattva, you need a deep aspiration: the aspiration to transform yourself and to help others transform themselves. You *need* to have an aspiration. And yet you can also practice aimlessness (or "businesslessness," as Master Linji called it). Aimlessness means "don't put something in front of you and try to attain it." Another way to understand it is like this: you are already what you want to become. Don't underestimate yourself. In this very moment, everything is here. You don't need to seek anymore. Everything is already here. Everything you're looking for, whether it's peace, well-being, happiness, or love—they are all already here. There's no need to keep searching.

The present moment contains the past and the future. If you know how to live your aspiration deeply in the present moment, you touch eternity. In Buddhism, the means and the ends should be identical. There is no path leading to happiness; the path is happiness itself. There is no practice that leads to healing and transformation; the practice should be healing and transformation itself. You have a deep aspiration but you can also practice aimlessness. It's perfectly possible.

Do You Dare to Dream?

—T.D.

As an engaged meditator, the energy of aspiration is one of the most essential elements of our path. We can ask ourselves, "What state is my life's aspiration in?" If it's a bit depleted, we can rekindle it. If it's still only a vague idea, we can crystallize it. If we found it before, but over time we've slowly let it go, we have to get it back. If it's buried deep, we can skillfully find a way for it to reveal itself.

There's sometimes a misunderstanding that the acceptance and equanimity of a meditator leads to a kind of serene or cold indifference in the face of the world's suffering, but this is only the case if we're using meditation to cover up the truth. In fact, the art of meditation is to *reveal* the truth and the path forward. Once we can see and understand how the world's suffering has come to be, right away we can also see how to transform it. There's a *dynamism* in the meditator's understanding of the world. Insight and understanding suffering gives rise to compassion and a deep wish to protect and nurture life. As a student of the path of bodhisattvas, our aspiration is a seed of vitality, a life force sustaining our journey.

There's a difference between aspiration and ambition. Making money, achieving a certain success, influence, position, or status are only external markers that society tells us we should realize. We may even be pursuing them without knowing it. Aspiration is much deeper than that. Aspiration is about what we really want to contribute to the world in this lifetime.

I remember when I was a young journalist, someone said to me that, if I continued working hard, one day I might be able to be a program "Editor" (capital E). It was a big word, and it sounded impressive. But something about it snagged, and I kept thinking about it. "If I keep going, one day . . ."

At the time, I had a practice of reciting the Five Remembrances before I went to sleep: reminding myself silently, breath by breath, that I am of the nature to get sick, to get old, to die, to be separated from those I love, and there is nothing I can do to escape that fact. The last line is to remember that I can take nothing with me when I die; my actions of body, speech, and mind are my only continuation. I began to realize that dying is a very private moment of reckoning. I will want to know: Have I been true to myself? Have I done my best to live well, to do what I wanted to do with this one, precious life?

One day it suddenly struck me that, if I kept going on the same track, my gravestone would say "Natasha Phillips, BBC Editor." I immediately realized I didn't want that. The next thought was "I'd rather die than have that on my gravestone!"

Many extraordinary people have contributed wonderful things to society as BBC Editors, but I saw that for me it was not the right path. The next realization was that, no matter what the gravestone says, on my deathbed I'll be confronted—above all—with my own relationship with myself. Will I have done what is most important to me? How can I not betray this chance to be alive, to heal and transform, not only my own junk, but also the junk of my ancestors and culture? How do I need to live my minutes and hours so I can be at peace on my deathbed? It was a new ground on which to stand to make decisions about my life. I had a deeper sense of what "a life lived well" might feel like, from the inside, and it gave me a wellspring of courage to walk a less-trodden path.

Tending the Fire

Being clear about how we want to live, and nurturing a strong aspiration, can be a powerful antidote to despair. Without it, Thay says, "we go insane" confronting so much suffering in the world. But it's not always easy to keep the flame alive.

In the first few years after ordaining as a monastic I was very active helping Thay with the community's engaged projects. But I hadn't yet learned how to find balance and at one point became exhausted and depleted. It was then that the seeds of despair came up. I realized I had to recalibrate and asked the community's permission to be released from computer work and organizing and to serve in the vegetable garden for one year. I spent my childhood in the countryside on a farm, and it was comforting to be back in the simplicity and calm of the mud, compost, and greens. But the sorrow was still there and, although the earth's softness and coolness were healing, the dark clouds in my heart were slow in clearing.

One day, a monk—Brother Spirit—brought me a message: "Thay asked me to tell you that he wants you to organize a retreat for journalists next spring." My immediate response was frustration. I had become a monastic in order *not* to be with journalists and to heal. My path had hardly begun! I wanted to claim my right to a simple life, a life close to nature and far from computers. Thay taught us about the importance of keeping balance and recognizing our limits. So, I sent back a defiant response: "No. Tell Thay that working on a computer is too stressful for me right now. I can practice mindfulness much better in the garden. The community has already given me permission to work in the garden for a year." Brother Spirit relayed the message to Thay, who just smiled. With a kind of fierce glee he brought down his Zen sword: "It's the same. Tell her working on the computer or planting lettuce: it's the same." I got the message, disagreed, and went back to my compost.

A week later, Brother Spirit had a new message. He gave me a draft of Thay's article "Intimate Conversation with Mother Earth," saying, "Thay is asking you to edit this. And he wants you to organize a press conference to release it." My stubbornness wanted to say no but my heart said yes. So, I asked some other monastics to help, and we did it together, and we eventually published Thay's beautiful writing as a book, *Love Letter to the Earth*. Conditions were not yet sufficient for the press conference. But, about two years later, we were able to organize a mindfulness event at the Dart Center for Journalism and Trauma in New York. It was intense and rewarding. Thay's arrow wasn't far off the mark; it just took time for the fire to catch.

Healing from burnout is complicated, and each one of us has to find our own path through it. We need time in nature, immersed simply in the present moment, soaking up the wonders; we need time with those we love and care about as well as time to take care of body and mind and to heal deep pains, regrets, and sorrows. We need to sleep, to cry, and to laugh. But what I learned from this

experience with Thay is that one of the most important things we can do is to keep the fire of aspiration alive; it's our most essential fuel. And, when it gets diminished, we need good friends on the path, loved ones and mentors, to help us keep moving and to remind us what we'd most like to do with our life.

One of my mentors once described "aspiration" and "mindfulness" as two wings of a beautiful bird soaring up high through a storm. We need both wings to ride the winds. We need the fire of determination, endurance, and non-fear—the qualities of Bodhisattva Ksitigarbha—and we also need the balancing strength of our mindful breathing and walking, eating well, sleeping well, relaxing, exercising, and being present for ourselves and taking care of whatever is coming up inside, right while we work to serve.

Guard Your Mind

Aspiration, or "volition," is one of the Four Nutriments. Another kind of nutriment is what we call in Buddhism "sense impressions"—what we hear, see, touch, and smell. When we watch movies and TV series, we consume. When we go online, we consume. When we read a book or magazine or listen to music, we consume. Even when we have a conversation, we are consuming that conversation. What another person says may be full of hatred and despair and, when we bring these things into our body and mind, it can be very toxic. The news can have a lot of anger, fear, anxiety, and hatred in it.

We want to have some kind of excitement, and we pick up our phone, our laptop, or a book or magazine, expecting to get it. We're looking for images and sounds that take us away from the discomfort we're experiencing in the present moment and cover up the suffering inside. When we reach for stimulation, it's not exactly because we *need* these things but we're doing anything we can to avoid encountering ourselves. And we can get addicted to these things, and yet never get the kind of fulfillment we need. We need love, we need peace, and it's only because we don't yet know how to generate that love and peace inside, that we are looking for it outside of ourselves.

The Buddha advised us to go home without fear, to breathe and walk mindfully to develop enough energy of mindfulness, concentration, insight, and courage to be able to take care of the loneliness and suffering that is there. With some training, you can bring your mind back to your body to reunite body and mind and create moments that can bring you love and joy. When you are truly there, you look outside and you realize that the rain is there, and it is wonderful; the trees are there, and they are beautiful; the air is surprisingly fresh. Breathing in and out you encounter the wonders of the present moment.

I remember, in the early 1970s, I was representing the Vietnamese Buddhist Peace Delegation at the Paris Peace Talks. We had so many things to worry about. The bombs were falling every day and so many people were dying. My mind was concentrated on how to help stop the war, how to stop the killing. And I did not have time to get in touch with the wonders of life that can be refreshing and healing. And so I didn't get the nutriments I needed. Sister Chan Khong was there as part of the delegation assisting me, and one day she prepared a basket of fresh, fragrant herbs. In Vietnam every meal is served with the leaves of fresh herbs. And I marveled. I had not had time to think of things like fragrant herbs and the wonders of life. That moment taught me a lot: that I shouldn't allow myself to be drowned in the work and be completely immersed in it. I should set aside time to live, to get in touch with the refreshing and healing elements within and around me. That day, I was able to restore my balance thanks to a plate of fragrant herbs presented to me by Sister Chan Khong.

Those of us who are activists are always eager to succeed in our efforts to help the world, but, if we don't maintain a kind of balance between our work and nourishing ourselves, we won't be able to go very far. That is why practicing walking meditation, mindful breathing, and getting in touch with the refreshing and healing elements in and around us is crucial for our survival.

Path of Heroes

Don't wait until tomorrow to cut through your afflictions. You must do it today. Sometimes we hesitate, and we continue to be entangled in our difficulties month after month, year after year. We can't get out of them. The warrior in us really wants to be free, the meditator in us wants to transcend the situation, and yet we allow our suffering to keep dragging us down. We want to put an end to the situation, but the warrior in us has not come into action, and that's why the meditator in us feels imprisoned.

Whenever we crave something—whether it's food, alcohol, or sensual pleasures—we can easily become a *victim* of our craving and lose our freedom. The objects of our craving are deceptively tempting, but there is a hidden "hook" that catches us. The way out is to recognize the hook so you can set yourself free. The warrior uses the sword of insight to cut through all afflictions and craving, and the warrior has to do it now. The warrior declares, "I can't wait. I must be free."

Whatever it is that is tempting you, whatever it is that has taken control of your life, whatever habit it is that has entangled you and made you suffer so much already—use the sword of wisdom to break free. Right now, today, straightaway, you make the determination not to go in that direction anymore. It's now or never. Give the warrior in you a chance to take action, to take out the sword and cut yourself free. Not tonight, not tomorrow, but right in this instant.

If you want peace, peace will be there right away. You declare, "I want to be free. I refuse to be dependent. I refuse to be a slave"—to the object of your craving, whatever it is. You want to be free. You want peace. The question is whether you want it enough. That's the key. If you're not free yet, if you haven't got the freedom, peace, and healing you're looking for, it's because you don't want it enough yet. You have to really want it, as though your life depends on it. Your determination to be free comes from your awakening. You see you've suffered too long already. You've had enough. You want out. You don't want to continue like that even one more day. That's the awakening. And from that place of awakening you resolve to be free.

Where's Your Horse Going?

There is a Zen story about a man on a horse galloping very quickly. At the crossroads a friend of his shouted, "Where are you going?" And the man replied, "I don't know. Ask the

ZEN AND THE ART OF SAVING THE PLANET

horse!" And that is the situation of humanity right now: in our times that horse is technology. It is carrying us off and it's out of control.

Early on, Google had the motto "Don't be evil." Is that realistic? Is it possible? Can you make a lot of money without being evil? That's what they're trying to do but, so far, not very successfully. Technology is making us alienated from ourselves, from our family, and also from nature—and yet nature has the power to heal and to nourish. But we spend so much time with our computer that we are no longer there for ourselves, our family, or Mother Earth. And that means civilization is going in a wrong direction. The way we are making money may not kill anyone, may not rob anyone, but it is costing us our life, our happiness, and the life and happiness of our loved ones and Mother Earth.

The truth is that, instead of using the insights of science and technology to save the planet, the market is using technology to satisfy our desires and exploit the Earth further. The problem is that we're using technology mostly to satisfy our cravings and take us out of the present moment. We *do* have enough technology to save us and the Earth, but the willingness to make use of it is not yet there. The question is, How can technology become a force for integration rather than destruction? How can tech companies innovate in such a way as to help people take care of themselves, take care of their workers, and take care of the Earth?

Beware the Embargo

We can consider our "store consciousness" as a kind of basement and "mind consciousness" as the living room. We have the tendency to put anything we don't like in the basement; we want to keep the living room beautiful. The same is true with our suffering. But the blocks of suffering in you don't want to stay in the basement and, if they become too strong, they push the door open and settle in the living room without your invitation. Especially during the night, when you have no means of control, they push the door and go up into mind consciousness.

During the day also, violence, craving, hate, and anger may be pushing hard because you have allowed them to grow through your way of consuming. You may try to resist. You lock the door. You set up a kind of embargo between the living room and the basement and you repress what is down there. And how do you repress? You try to fill up mind consciousness by consuming. You feel uncomfortable, restless, you feel something is coming up, and so you switch on some music, you pick up the phone, you turn on the TV, you head out to go somewhere. You do everything to keep the living room occupied so that these blocks of pain have no chance to come up.

But many of the things we consume in these moments may also contain a lot of toxins, and while you are consuming them, the poisons of craving, hate, and violence fall down into store consciousness and they only make the blocks of suffering bigger. This is a very dangerous situation. So, the first step is to stop watering these seeds to prevent them from growing. You have to establish a strategy of mindful consuming with your loved ones and friends—a strategy to support and protect each other.

Suppose we speak of depression. Depression does not come by itself from nowhere. If we look deeply into depression, we can see its roots. Even a depression needs food to stay alive. And so, we must have lived and consumed in such a way over the previous months that has made the depression possible in the present moment. Perhaps we have suppressed our suffering with our way of consuming and created a situation of bad circulation in our consciousness. With mindfulness, you respond differently. You can allow the pain to come up, you recognize it, embrace it tenderly, and look deeply into it. Embraced with the energy of mindfulness, concentration, and compassion, your pain will go down, losing some of its strength. And, the next time it comes up, you allow it to come up because you know how to handle it. After a few weeks of practicing like that, you restore a state of good circulation in your consciousness. The practice of mindfulness can be very healing.

A Warrior's Strategy to Guard Our Mind

—T.D.

The concept Thay is proposing here is relatively simple: to begin to see everything we read, watch, and listen to as food. The tricky part is then transforming our habits, which are also the habits of our culture, ancestors, and civilization. Mindfulness gives us a chance to be vigilant and alert and to feel how our body and mind are responding to the input. How do we feel while scrolling, and how do we feel after? How do we feel about the films or series we're watching, especially once we switch it off? What's the imprint? Has it triggered tension, fear, agitation, and loneliness? Or joy, fulfillment, connection, and understanding?

Screens provide us with many good things: laughter, inspiration, education, entertainment. TV, movies, and so on are not of themselves bad; they have their benefits. The challenge is to train ourselves not to drown in them. In Plum Village during our monastic retreats, we have a movie night. And, in the summer, we often watch the soccer World Cup final, together with hundreds of people, in the meditation hall.

There are no absolutes here, no "rights" or "wrongs"—mindful consuming is an art. It's about content: Is what we're consuming polluting our minds with fear, violence, or anger? It's about time: How much is enough? And it's about what our screens and earphones are taking us away from: the presence of a loved one, time out in nature, or simply the chance to be present for ourselves and the feelings coming up inside. For everyone it's different. But, the stronger our practice of mindfulness gets, the more clearly we see, and the more free will and options we have. And, as we learn how to handle the most acute and difficult feelings, the less we're afraid of what we confront in ourselves once we switch off the screens.

Taking control and having the freedom to choose what we put in our mind is not easy: there are sophisticated algorithms and supercomputers stacked up against us, explicitly designed to game our preferences and to profit from gripping our attention. In 2013, a few of us monastics accompanied Thay when he met senior executives and engineers in a boardroom at the Google headquarters in Mountain View, California.

Thay had just given a talk to hundreds of employees about how important it is to create the kind of apps and devices that can actively help us suffer less and take care of our body, feelings, and relationships. The engineers had questions. They wanted to know what Thay would do if he were in their position; they wanted to know where to draw the ethical boundaries.

Thay listened deeply to their concerns. He was patient and attentive, generous and encouraging but also firm: if we can help reduce suffering, we should do it. One of the engineers who was there that day went on to establish the Center for Humane Technology, a non-governmental organization working to promote a digital world that supports human well-being rather than exploits human weakness. Individually and collectively, it's up to us to resist a digital future that—without us even realizing it—monetizes our attention, radicalizes our views, and fuels an economy of clickbait and craving.

The question is, How? We could do with a strategy: deciding how many hours a week we'll watch films and TV, play games, read the news, or scroll social media, or using apps and blockers to help guard our attention so we can stick to our resolution. Can we commit to putting away our phones when we're with loved ones or leave our phone in a different room when we go to bed? These are simple decisions and yet they're hard. Self-mastery starts small.

We also need a plan for how we'll take care of our mind when we're *not* consuming. How will we embrace our loneliness, sorrow, or despair? How will we nourish joy and connection? How will we relax? I have my own short list for what works. When the seed of sorrow comes up, I do my best to get outside and bring my full attention to the present moment—the sounds, sights, smells, taste, and touch of the miracle of this thing we call life.

When I'm restless or anxious I can feel it in my bones and take a few minutes to practice a body-scan relaxation, either sitting or lying down. When anger is triggered, I try to go for a walk as soon as possible. Generating compassion helps—it's the most powerful antidote for rage. And sometimes, just behind our anger, we can find the hidden hurt and fear, triggered in the nanosecond just before, and we can tend to that.

When I have a disgruntled state of mind that just won't budge, I train myself to "change the channel," to switch track, shift topic, and bring up something more positive. This is what's called in Buddhism cultivating "appropriate attention." Sometimes chang-

ing the object of our attention is the best thing we can do to restore balance; at other times it can be avoidance or escapism.
The trick, the art, is to learn to discern what our mind needs—
what would be most healthy—in any given moment or context.
I've also learned to never underestimate the importance of good
friends: people just to hang out with, talk, laugh, play, and cry
with. The pandemic made this essential resource even more
scarce and precious. Simply being together with other human
beings can be a deep and fulfilling source of nourishment, and
this "collective consciousness" is the next of the Four Nutriments
in Buddhism.

Collective Consciousness Is Food

So, there is the nutriment of "volition" and the nutriment of "sense impressions." And the third kind of nutriment the Buddha spoke about is *consciousness*. We can speak of *individual* consciousness and *collective* consciousness. The individual is made of the collective and the collective is made of the individual; they inter-are. Our individual consciousness reflects the collective. For example, the fear and anger in us are *individual* but they also reflect somehow the fear and anger of society. Or suppose you think something is beautiful. It's not just because it is beautiful in itself but because it's considered by the collective consciousness to be beautiful. The individual consciousness and collective consciousness that we consume every day are *very real*.

Thoughts, feelings, and states of mind can all be considered a kind of food entering our body and mind. If we sit there and allow suffering or sadness to come up, and we chew over it, again and again, like ruminant animals chew the cud, consciousness can become an unwholesome kind of food. But with the energy of mindfulness, we can train to handle our thinking with appropriate attention, in a way that leads to understanding, compassion, and freedom. We can choose *when* it's appropriate to bring our attention to different kinds of thoughts. Our consciousness is a source of food, and with mindfulness we can choose which food we know will nourish us and help us grow.

If we see a friend who is sunk in thought and carried away by their thinking, and whose face looks anxious and sad, we know they may be drowning in their thinking or in a feeling of pain and sorrow. This is "*in*appropriate attention." In such moments, we have to help. We put a hand on our friend's shoulder and say, "A penny for your thoughts! It's a beautiful day. Let's go for a walk." We help pull our friend out of consuming unwholesome consciousness food. We don't let them sit there ruminating. Anyone who can benefit from a zone of positive collective energy with good friends will start to feel better; and day by day, they'll feel nourished and transformed.

The same is also true with negative energy in the collective consciousness. When fear and anger become collective, it can be extremely dangerous. You can be as afraid and alarmed as everyone else; and easily swept away by the collective energy. That is why it is very important to choose an environment where you'll be influenced by a healthy, clear collective consciousness. Many of us are influenced by thinking around us. For example, during the Iraq War in 2004, 80 percent of Americans believed the war was just, whereas in the UK only 35 percent thought so. A whole country can lock itself into one notion, one idea, one feeling. The mass media, the military-industrial complex, all these elements may create a prison for us, so we continue to think in the same way, to see in the same way, and to act in the same way.

Who can help us get out of that prison of views, of feelings, the prison of collective consciousness? You. Whether you are an artist, a writer, a journalist, a filmmaker, an activist, or a practitioner, you should generate your *own* insights. You help the Buddha in you to manifest, and you express yourself in the light of truth. And, even if the majority has not seen the truth that you have, you are courageous enough to continue. The minority who have the truth with them can transform the whole situation.

Eating with Non-Violence

The fourth kind of nutriment the Buddha spoke about is *food*: breakfast, lunch, and dinner—in fact, anything we consume by way of our mouth. What we eat is very important. Tell me *what* you eat and I will tell you who you are. Tell me *where* you eat and I will tell you who you are.

The Buddha said that we should eat in such a way that keeps our compassion alive. UNICEF reports that, every year, three million children die from hunger and malnutrition. They are our own sons and daughters. And, if we overeat, it is as though we are eating them. Our way of eating and producing food can be very violent. We are eating our children; we are eating our descendants. We're eating up the Earth.

Mindfulness helps us be aware of what is going on. The meat industry has devastated our planet. Forests are being destroyed to create grazing land for cattle or to grow crops to feed them. The world's cattle alone consume a quantity of food equivalent to the caloric needs of 8.7 billion people. It takes a hundred times more water to produce a single pound of meat than it does to produce a pound of grain.

Urgent action must be taken at the individual and collective levels. Not eating meat is a powerful way to help our planet survive. Simply by eating vegetarian, you can preserve water, reduce pollution, prevent deforestation, and protect wildlife from extinction. If we stop consuming, they will stop producing.

It's your awareness of suffering that naturally makes you determined to consume non-violently—it's not because someone forces you. You do it out of awareness, mindfulness, and compassion. It's a way of expressing love and gratitude to the Earth. And you can have peace, joy, and happiness right away. Our daily life has to express our awakening. Only by consuming mindfully can we keep our compassion alive and ensure our planet can have a future.

You can be very happy following a vegetarian diet, and you can do it without judgment. You stay tolerant; you don't want to impose your ideas on others. You allow them to be themselves. You should not talk too much about it but simply invite others to enjoy delicious vegetarian dishes with you. There will always be people who continue to eat meat or drink a lot of alcohol, but we need 50 percent of humanity to volunteer to create balance. Eating is just one part of the practice, and as others see your peace, joy, and tolerance, they will begin to appreciate a non-violent way of eating. If our society can practice mindful consumption, we will be able to heal ourselves, heal our society, and also heal the planet.

In the Buddhist tradition there is a verse reminding every monastic to train, in their interactions with the world, to be as gentle and mindful as a bee visiting a flower. Bees feed on a flower's sweet nectar but without destroying the flower's fragrance and beauty. We are all children of the Earth, and we can make the most of the Earth and benefit from its beauty, but in such a way that we respect the Earth, just as a bee respects the flower.

Perhaps businesspeople and politicians need to contemplate this because we have destroyed the Earth, we have destroyed the flowers, waters, and mountains with our greed. We should take only what we need and behave in such a way that the beauties and flowers of the planet remain intact. The teachings on the Four Nutriments help us understand how not to destroy ourselves, our community, or our planet, and how to nourish and protect ourselves and our environment.

Learning the Art
of Nourishing and Healing

—T.D.

Courage and a radical honesty help us to see clearly the effect of our way of consuming on ourselves and on the planet. Is the pace of our growth economy sustainable? Am I the master of my attention? What's the true cost of a cheap T-shirt, a glass of liquor, or a pound of beef? The truth may sting, but it can wake us up. That insight will help us transform our habits, so we can find new ways to sustain our aspiration and find healthy fuel for our body and mind.

We need the determination and strength of a warrior to cut through and claim our freedom, but we also need the patience and kindness of a meditator and the openness, tolerance, and creativity of an artist. As Thay says, "Don't make yourself into a battlefield!" The world doesn't need any more fanatics. If a habit is hard to shift, it's likely to have been transmitted to us through several generations or held in place by society, culture, and our context or environment. We can discover a lot about ourselves and our ancestors as we begin to make changes to align our choices with our values.

Here is the text of the next of the Five Mindfulness Trainings that Thay wrote to guide us as we navigate our own journey toward a more mindful way of consuming. You may find it stark, confront-

ing, or challenging to read. If so, that's okay. The text is meant to be provocative, like a Zen master asking, "Are you sure?" You may like to read the text slowly and reflect on the ways in which it speaks to your life right now, or challenges it. Each reading may bring up a different response. Which insights, questions, or reactions does it bring up for you today?

 ✳t.d.

The Mindfulness Training
on Nourishment and Healing

Aware of the suffering caused by unmindful consumption, I am committed to cultivating good health, both physical and mental, for myself, my family, and my society by practicing mindful eating, drinking, and consuming. I will practice looking deeply into how I consume the Four Kinds of Nutriments, namely, edible foods, sense impressions, volition, and consciousness. I am determined not to gamble or to use alcohol, drugs, or any other products that contain toxins, such as certain websites, electronic games, TV programs, films, magazines, books, and conversations. I will practice coming back to the present moment to be in touch with the refreshing, healing, and nourishing elements in me and around me, not letting regrets and sorrow drag me back into the past nor letting anxieties, fear, or craving pull me out of the present moment. I am determined not to try to cover up loneliness, anxiety, or other suffering by losing myself in consumption. I will contemplate interbeing and consume in a way that preserves peace, joy, and well-being in my body and consciousness and in the collective body and consciousness of my family, my society, and the Earth.

BRAVE DIALOGUE:
THE POWER
OF LISTENING

In True Dialogue Both Sides
Are Willing to Change

If you want to save the planet and transform society, you need brotherhood and sisterhood; you need togetherness. Whenever we speak about the environment, or peace and social justice, we usually speak of non-violent actions or technological solutions, and we forget that the element of collaboration is crucial. Without it, we cannot do anything; we cannot save our planet. Technical solutions have to be supported by togetherness, understanding, and compassion.

In order to collaborate, we need to know how to listen deeply and how to speak skillfully, how to restore communication, and how to make communication easier so we can communicate with ourselves and with each other. We may have a lot of goodwill to come together, analyze the situation, draw up a plan, and take action. But, if we can't agree, if we only fight with each other, our organization will disintegrate. If we don't know how to help each other, or how to listen, we get angry and become divided.

Restoring communication is an urgent practice. With good communication, harmony, understanding, and compassion become possible between individuals, different groups, and even nations. Our leaders need to be able to come together to talk about the danger that our planet is facing. And the practice of deep listening to each other, using skillful, compassionate speech to convey insights and ideas, is so important for opposing parties to establish real human relationship and understanding.

Political leaders and representatives need to train themselves in the art of deep listening: listening to their own people, to the suffering in their own country, and to the suffering of other nations. Many people feel their suffering is not heard and understood. Our societies are deeply divided. We are killing each other, and there is fear, anger, discrimination, and despair because—as a species—there is not enough communication among us. We're not only killing *other* species but we're killing *ourselves* as a species. And that is why we need to learn how to listen deeply so each one of us can contribute our listening and compassion.

When we hear words of blame and judgment from others, it may touch off our own irritation, anger, and frustration. And so, we need more than just the *intention* to listen: we need *training* in how to listen. When we can open the door of our hearts and restore communication in our close relationships, we'll be able to do the same at work, across society, and among different political parties and nations.

As we listen deeply to the other side, we begin to recognize not only their wrong perceptions but also *our own*. That is why mindful dialogue and communication is so crucial. Practicing deep listening and loving speech can help remove the wrong perceptions that are at the foundation of fear, hatred, and violence. It is my deepest hope that our political leaders can make use of such instruments to bring peace to themselves and to the world.

Bridge the Chasm

There is a special bodhisattva in the Lotus Sutra known as "Earth Holder," Dharanimdara. The name means "the one who holds the Earth"—the one who protects and preserves. They are really needed in our times. Earth Holder Bodhisattva contributes the energy of holding life together, sustaining life. Their task is to further communication, connection, between humans and other species and to protect the Earth and the environment. Their role is as a kind of engineer or architect whose task is to create space that can embrace everyone, to build bridges to cross from one side to the other, and to build a road, a way, so we can reach those we love. We should recognize the presence of the Earth Holder Bodhisattva within us and around us. We, too, can create space that includes everyone. We can help people to restore communication and build a bridge from heart to heart.

I came to the West in 1966 above all to speak out against the Vietnam War, but together with other friends in the peace movement we also meditated on the wider situation of our planet and focused our attention on the Earth. After a lot of deep looking, we set up an organization called Dai Dong to create awareness of the need for a *transnational* community of humankind, a community that transcends the boundaries of nations. Đại means "great" and đồng means "togetherness": an organization for Great Togetherness.

In 1970 Dai Dong convened a meeting of scientists in France, and we released the Menton Statement, a "message to our three billion and a half neighbors on planet Earth," signed by over three thousand scientists. We spoke out about environmental deterioration, depletion of natural resources, population overcrowding, and hunger. In 1972, when the UN organized the first environmental conference in Stockholm, we organized a parallel alternative conference, not as governments but as the people, acting in the spirit of Great Togetherness.

The Secretary-General of the UN at the time, U Thant, endorsed our efforts, saying, "This global concern in the face of a grave common danger—which carries the seeds of extinction for the human species—may well prove to be the elusive force which can bind man together." If all the conflicting parties in the world could be directly exposed to the truth and given full information about the situation of the planet, then we would stop our disputes. We would solve our conflicts quickly in order to do something. But, if nations are still concerned about things that aren't important, it's because the full awareness of the truth is not yet there. As soon as we accept the truth that, if we continue like this, we cannot avoid the end of our civilization, we'll have the strength and awakening we need to bring us together, to overcome anger, division, hatred, and discrimination, and to see clearly what we need to do.

How to Listen

Even though we have more sophisticated technology for communication than ever before, communication has become very difficult. Many of us cannot even listen to ourselves. To listen is first of all to be fully present and not distracted. It means to be present for ourselves; to listen to ourselves and, with mindful breathing, to restore our peace and freshness. This is the quality of presence we offer to the person in front of us so we can listen to hear what is being said and what is being left unsaid.

There is an art to listening. Mindfulness is always mindfulness of something, and in many cases of listening, we are practicing *mindfulness of compassion*. You keep alive the insight that you are listening with only one purpose: to give the other person a chance to speak out and suffer less. And you practice mindful breathing while listening to keep that intention alive. And, even if the other person says things that are not true, or that are full of wrong perceptions, bitterness, or blaming, you can continue to listen with compassion because you are practicing mindfulness of compassion. Compassion protects you and prevents what the other person is saying from triggering irritation and anger in you. One hour of listening like that can help the other person suffer much less and can help restore communication.

But, if you find you can't listen, and you turn away, it's not necessarily because you don't have compassion but because you have not yet been able to transform your own suffering inside. I remember hearing about one time that Oprah Winfrey was interviewing a woman who was talking about her suffering of being abused. She nearly broke down as the interviewer and was so distressed she wanted to stop the camera. It was because she had gone through the same kind of suffering; she had been abused as a child herself, and she had not had an opportunity to transform that suffering, so, listening to the suffering of the other lady, she nearly broke down. So, the truth is that, we have to learn how to transform our own suffering so we can be there for those who are suffering and need our help.

I believe that teachers and professors also need to devote some time to listening to the suffering of their students. It's not a waste of time because, if students suffer too much, it's difficult for them to learn. And so, helping them suffer less is essential to the work of teaching and transmitting knowledge. Once a teacher has listened to their own suffering, they can listen to the suffering of their students, and help them suffer less, within even just an hour of listening. And, in return, students can then listen to the suffering of their teachers because educators also have their own suffering, and, when they can share that skillfully with their students, the communication will become much easier. The atmosphere will be transformed and the work of teaching and learning will become much more enjoyable.

In every organization there should be people who know the art of compassionate listening and using loving speech. And their job is just to go to each person to sit down and listen. They may say something like this: "Dear friend, please tell us how you are doing, please help us understand the challenges and difficulties you are facing in your work, in yourself, and in your family." The listener should be someone who has already listened to themselves and to their own family before they can do the job of helping others speak out and suffer less. With good communication among its members and everyone sharing the same intention and direction, any organization can be transformed into a community that can be an instrument for social change.

There are times, also, when we need to listen to ourselves as a nation. There is a lot of suffering within the country; there is injustice, discrimination, and anger, and many people feel their suffering is not yet heard and understood. We need to identify those of us who have the power of listening to come together and help us practice as a nation to listen to those who feel they are victims of discrimination and injustice. And we can create the kind of environment where they feel safe to express themselves, even if it takes several days or weeks to get enough courage to share everything that is in their heart. And, once we've been able to listen to each other as a country, we'll be able to listen to those in other countries.

Bodhisattva of Compassion

Avalokiteshvara is the bodhisattva who knows how to listen deeply to the cries of the world. All of us—and all species— express ourselves in different ways; no matter how we express ourselves, Bodhisattva Avalokiteshvara can always understand. If that person is a child, who doesn't have enough words to express themselves, the bodhisattva understands. If that person expresses themselves in words or bodily language, the bodhisattva also understands.

We need to learn how to listen; we need to train. If what the other person is saying is difficult to hear, you may have an urge to stop and correct them because it's painful to hear and touches your own suffering. But we do our best to refrain from interrupting them. It's not important whether what they are saying is right or wrong. What is important is to give them a chance to unburden themselves. Listening with compassion for their suffering is the only way to help. The other person needs someone to listen to them, and you may be the first person in their life who's been able to listen and give them a chance to empty their heart. It is a very deep practice, and it takes training.

You tell yourself, *They are suffering, and they need someone to listen, and I will be that person. I will play the role of the bodhisattva of deep listening, Avalokiteshvara. I'll be able to do it, if I remember to breathe in and out mindfully while listening and remind myself of one thing: I am listening with the sole intention of allowing them a chance to empty their heart. Whatever they say, even if it's wrong or full of accusation, blaming, and wrong perceptions, I'll still listen.* And that is called compassionate listening. It's very kind of you to sit and listen like that. You are playing the role of a bodhisattva. You are protected by the energy of compassion. This is what we have to train ourselves to be able to do.

It is possible to listen to suffering in such a way that we don't get sucked in. We all have the seeds of both anger and compassion. If you practice mindfulness effectively, perhaps the seed of compassion in you will be bigger than the seed of anger. And, if the seed of compassion is powerful enough, you'll be able to activate your compassion to protect you while listening, and you'll be able to protect the seed of anger from being triggered. In compassionate listening our intention is not to insulate ourselves or cut ourselves off from the other person. We listen to others' suffering with compassion, and take care to recognize, embrace, and transform our own wounds that may have been touched while listening. In Plum Village, after a session of deep listening, we often practice walking meditation outside to restore our peace, calm, and freshness.

At the same time, we have to know our limit. We should organize our life so we have balance and enough nourishment, peace, and joy so we can continue to offer compassionate listening to others. Some people have so much pain, hate, and violence in them, and they need to express it. And it's very hard for them to find someone who can sit and listen to them. So, when you come ready to listen to them, they may have the tendency to abuse your time and kindness and speak nonstop. And you don't know how long, how many days or years of listening will ever be enough to help them. They repeat the same things again and again.

To keep listening in such a situation is not intelligent deep listening. We have to find skillful ways to actively help them recognize, embrace, and deeply transform the suffering in them. Being listened to, they may be getting a little bit of relief, but it's not enough. So, we should find allies and together help them organize their life so they can cut whatever source of nutriment is feeding their suffering. Simply to keep listening, we may be destroying ourselves and the bodhisattva in ourselves, and that is not a good thing at all.

Deep Listening 101

—T.D.

Whenever we disagree with someone, it can take courage to hear them out, and, whenever we experience injustice of any kind and feel powerless, it takes immense spiritual strength not to fall victim to anger or hatred. How can we help build bridges in a fractured, polarized world?

Before visiting Plum Village for the first time, I didn't realize listening was learnable. I just figured that you either have the gift, or you don't—and I didn't. But gradually I discovered that, the more I was able to simply be still and listen to myself, the more space I had to listen to others; the more I listened to the skies and trees, the more I could listen to human beings. I got curious about people and the landscape of their hopes, fears, and dreams. The thing is, it's easy to mix up listening with trying to get our own point across, when in fact speaking and listening are two different things. Listening is a training, a practice. It's a gift we offer the other person, and it's a gift we offer ourselves: to expand our perspective and encounter the human being in front of us in a deep way.

In Plum Village we train to listen with our whole person, to be fully present for what's being said. The first trick is to follow our breathing while listening. Right away, we become an embodied listener. Paying attention to the extraordinary symphony of our breathing keeps us rooted in the present moment and helps us not get distracted by our own internal discourse. Breathing mindfully as we listen, we soon discover that our breathing contains within it the trace of our reactions. By taking care of our breathing, we have a chance to recognize, receive, and embrace any reactions right away as they arise.

There is a lot going on: there is the person in front of us and their words and there is our own body, breathing, and reactions. This is the second trick: to take care of the impression of the other person's suffering in the way it affects our breathing and our body. If tension arises, we release it on an out-breath. If our breathing gets uneven or short, we soften and gently release it. We don't repress any feelings that come up for us; we simply take note of them and embrace them, knowing we can always look into them later after we have finished listening.

The art of listening includes the art of not-interrupting, the third trick. When someone triggers us, or says something false, the first thing we might want to do is to interrupt, correct them, and explain why they're wrong. But, in deep, compassionate listening, our task is above all to allow the other person to say everything they have to say. It's our chance to hear what they really think, in their heart of hearts. And, if it's painful to listen, as Thay explains, we protect ourselves with the energy of compassion and remind ourselves that we're listening with just one purpose: to let them open up and speak out. We cultivate a genuine curiosity to understand their deepest fears and concerns.

This is the fourth trick: to keep our compassion alive during the whole time of listening. Mostly I do this by not focusing too much on the words. I find that the best way I can listen to difficult, bitter, or angry speech is to listen to the pain *behind* the words, to the feeling the person is trying to articulate, however clumsily they're doing it. Thay described the action of the bodhisattva of deep listening, Avalokiteshvara, as listening "so attentively that we will be able to hear what the other person is saying, and also what is being left unsaid." It's a struggle for all of us to put our feelings into words, even at the best of times, and especially at the worst, when we've been hurt or when our fear or anger has been triggered.

Sometimes, when I'm listening to someone who is agitated or angry, as I'm looking right at them, beyond their words, and following my breathing, I hold a question openly and silently in my heart: "What's really hurting here? What are you really trying to say?" There can be a huge chasm between what one person means to say and the words we actually hear. The energy of mindfulness can help us bridge the gap. This is the fifth trick: hearing what's being left unsaid.

Finally, we have to create the right kind of conditions for listening to be possible. Whether it's putting down cell phones, switching off the TV or music, or suggesting going outside for a walk or coffee—in any given situation there's always something we can do to create a slightly better context to be fully present.

At the same time, we need to be honest with ourselves: Are we really ready to listen? Do we have enough space inside? If we're not in the right frame of mind to listen, it is better to say so, and offer to listen deeply another time. We have the right to respect our own limits too. There's an interbeing between speaker and listener: when someone is really listening to us, when we really feel we'll be heard, suddenly it becomes possible to fully express what's in our heart. In the same way, when someone is not listening with genuine openness and compassion, we can feel it.

I've noticed that taking communication to a deeper level can be hardest in our closest relationships. And yet it can be very powerful to go for a walk with someone we care about and ask, "How are you really doing?" or "What's your deepest concern right now?" or even "Do I understand you enough?" Some say that, if you don't know how an extrovert is feeling, you haven't been listening, and, if you don't know how an introvert is feeling, you haven't asked. The strange thing is that sometimes the ones with the loudest voices are also those who feel the least heard.

When I was working as a young journalist, there was a colleague on the team who lived alone. Every day he burst through the office door with a toxic rant about the traffic, the breaking news of the day, or the politician he'd met down in the lobby. Most colleagues, in a very British way, would mechanically call out from their desk, without even looking up, "Good morning!" and get on with their work. I also tried this, but, as I was the most junior, they'd given me the desk nearest him. Ignoring was difficult. I remember one day the rant had been going on for more than ten minutes, and it was impossible to get any work done. He was so angry and cursing so much, I met my absolute limit. It was too toxic. But he was my senior, so I couldn't just ask him to be quiet, and I had work to do, so I couldn't just walk out. But then I thought it might be a good idea to try listening.

So, I swung my chair around to face him, and started listening 100 percent, following my breathing, and calmly, openly, looking straight at him. First, I realized he was taken aback that anyone was actually listening. And then, within a few seconds, all I could see on his face was the loneliness and frustration, echoing in the dark. I honestly felt compassion. So, I sat and breathed and listened and breathed. And within just a couple of minutes he ran out of steam, saying, "What are you listening for? Get back to work!" And then, in a change of tone, "I'm getting a cup of tea. You want one?" And he disappeared off to the kitchen. From then on, all I had to do when he came in ranting was swing my chair around, and he quieted down. Sometimes even if we think we want to be heard, we're not always ready for it.

 *t.d.

Mastering Anger

Some people see anger as energy and feel we should make use of that energy to fight for justice and social equality. Of course, anger is very powerful, but the question is whether you can *control* it. When you are angry, you're not very lucid, and you risk creating a lot of damage, both in yourself and in the world. But, if you know how to transform anger into compassion, you still have a very powerful source of energy. And with that energy of compassion, people can be ready to die to save the lives of others; they have nothing to fear, just like a mother who sacrifices her own life to save her child. Compassion is a better energy than anger, just as solar energy is better than nuclear. So, the practice of mindfulness here is not to fight or suppress the anger but to recognize, embrace, and gradually transform it into compassion. We need some training to do so.

Learning how to handle our strong emotions when they arise is essential to succeeding in the practice of loving speech. Compassion is a kind of *antidote* for anger; the two are somehow related. As soon as compassion is there in your heart, you're no longer angry; you can restore communication and reconcile. You can communicate with yourself much more easily; you can understand yourself, and you can communicate with others more easily. Anger, on the other hand, blocks communication.

I always advise couples that, whenever you get angry with each other, you should return to your breathing and mindful walking, embrace your anger, and look deeply into it. Within just a few minutes, or a few hours, you may be able to transform your anger. But, if you can't, you have to let the other person know. And you try to say in a calm way: "I want to let you know that I'm suffering. I honestly don't know why you did what you did or said what you said." And, if you're not calm enough to say it, you write it down on a piece of paper.

The second thing you can say is "Please know I'm doing my best." It means "I'm trying not to say anything or do anything out of anger because I know that, if I do, it will only create more suffering. I am doing my best to embrace my anger, and look deeply into it." You let them know that you are trying to find out whether the anger may also come from your *own* misunderstanding, wrong perceptions, or lack of awareness.

The third thing you might like to say is "I need your help." Usually when we're angry, we want to do the opposite. We say, "I don't need you anyway. Leave me alone." But, if we can say, "I need your help," it means "I need your support to overcome my anger." The quality of our being is very important. You are free to share what's in your heart. But you have to say it in such a way that the other person can hear so they can really listen. If we are too aggressive or blaming, there's no point. Expressing yourself is an art. If you still have too much anger, too much energy of blaming and punishing, you'll only widen the division. And so, you write a note in such a way that it's a *real* invitation to restore true communication. You are ready to listen; you are ready to understand.

The Art of Not Hating

Very often in a conflict we believe that we can only be happy and peaceful if the other side simply didn't exist anymore. We may be motivated by the desire to annihilate, to destroy the other side, or even to lock them up. But looking deeply we know that, just as we have suffered, the other side has suffered too. We want a chance to live in peace, safety, and security, and we also want the other side to have a chance to live in peace, safety, and security too.

Once you are able to include the other side in your heart, once you give rise to that intention, you'll already begin to suffer less. And it becomes possible to ask them, "How can we best ensure your safety? How can we help you have peace and security and opportunity? Please tell us." As soon as you are able to ask those kinds of questions the situation can change very deeply, very quickly. But, first, there must be the change within your heart: the intention to *include* others, to give them a chance. With that intention, you suffer less right away; you no longer have the desire to eliminate.

So long as we see the other person as the enemy, we are

determined to win, to punish them. The more they suffer, the more we are pleased. But, with this way of thinking, we will surely fail. The Buddha teaches us that first we have to win *ourselves*, meaning we have to free ourselves from resentment, hatred, and wrong perceptions. We have to win our own mind first. Winning does not mean victory over those who cause us to suffer but victory over our own ignorance and resentment inside. We may have the impression that we're blameless and that all our suffering has been caused by the other person, the other side. But that's not true. We've been responsible for at least part of the suffering. And, if we look, we'll be able to see that. And, if we still can't see our part in it, we can ask others to show us.

Those we call "terrorists" have tremendous resentment and hatred. They have suffered, and they make use of the energy of their hatred to drive action. But this does not mean that those who do not consider themselves "terrorists" are without hatred. Which side does *not* have hatred? Which side does *not* have misunderstanding?

We may consider ourselves righteous, walking on the right path, without blame, without hatred. And we may consider the other group as a threat to civilization or global security. More than ever, we need to use the sword of understanding to free ourselves and others from labels. This side labels the other side, and vice versa, in order to resist one another, or even kill one another, in the name of God, democracy, freedom, or civilization.

At Plum Village in France, we have invited groups of Palestinians and Israelis to come and practice with us. It is always difficult in the beginning. When the two groups arrive, they cannot look at each other, they cannot talk to each other because both groups have a lot of anger, fear, and suspicion—a lot of suffering. So, during the first week we let them stay separate, and they practice mindful breathing and mindful walking, to help them to calm down their pain and embrace their pain. They're guided to get in touch with the wonders of life within themselves and around them to get the nourishment they need.

At the beginning of the second week, we train them in deep, compassionate listening and loving speech. One group is asked to speak about their suffering, while the other group practices deep, compassionate listening. With that practice, people suffer less right after the first session. When you listen like that, you realize that the other side have suffered exactly like people have on your side. For the first time, you see that they are also victims of the conflict, victims of wrong perceptions. Understanding and compassion are born in you, and you're no longer angry. You're able to see, for the first time, that they are just human beings like you. They have suffered very much like you have. So, mutual understanding and communication become possible. And you know that later, you, too, will have a chance to speak about your suffering, difficulties, and despair and the other group will listen. The practice of compassionate listening can remove a lot of anger, remove a lot of suspicion, and remove a lot of fear.

During the Vietnam War, America sent half a million soldiers to Vietnam, and they killed many Vietnamese people. They destroyed our villages and children. About fifty thousand American soldiers died in Vietnam, and many hundred thousands of them went back to America and became very sick and needed a lot of psychotherapy. My practice is not to hate these American soldiers who came to Vietnam because they are also victims of a policy that is neither intelligent nor compassionate. And, whenever we visit America to lead retreats, we offer retreats for American veterans to help them heal and begin their lives anew. Many have gone back to Vietnam to help repair the damage they have done in our country.

Forgiveness is possible. That is my own experience. I have undergone a lot of injustice and suffering. I have survived wars. People have committed injustice against me, my people, and my nation. Compassion and forgiveness are possible once we can *see* the suffering of those who've inflicted suffering on us. And, when we understand that suffering, we have compassion and we're able to forgive.

The quality of love embodied by the Buddha is possible for us to attain. It is possible to respond to hate with love. It is possible to respond to violence with compassion and non-violent action. With deep listening and loving speech, you can change people's way of thinking; you don't need to kill them. We don't need to kill terrorists. What makes them into terrorists is their hate, fear, and anger. But, if we sit and listen to them deeply, we can help them transform their anger and fear and they will stop being terrorists. You can only help remove wrong perceptions by dialogue: deep listening and compassionate speech.

Is It Possible to Work for Change Without Hating the "Other Side"?

—T.D.

When Dr. Martin Luther King Jr. marched in downtown Chicago against the Vietnam War, on March 25, 1967, a few months after he and Thay first met, he did so under a banner in English and Vietnamese that read "Men are not our enemies. If we kill men, with whom shall we live?" Thay and Dr. King shared the insight that it is not other *people* who are our enemies; instead, our enemies are anger, resentment, hatred, fear, and discrimination. It was by concentrating on this insight that Thay, Sister Chan Khong, and their young social workers in Vietnam were able to stay neutral and not take sides during the war.

The path of "not hating the enemy" is a profoundly spiritual practice. Dr. King once said that, if we have enough spirituality and ethics, we'll be able to replace hate with love, and "love the individual who does the evil deed, while hating the deed that person does." As meditators, we train in this way of seeing so we can confront injustice without hating. We all need a spiritual dimension in our lives, and that is why Thay says we shouldn't allow spirituality to be exclusively claimed by one particular side. And yet, sometimes, he says, that's exactly what happens: "One side monopolizes God: they hijack God in the direction of division, hatred, discrimination, and intolerance, and try to show that the other side goes against the will of God." And yet, what the world most needs now, Thay says, is "a god of compassion. We need a god of nondiscrimination, we need a god of tolerance, we need a god of love." Love allows us to go even further, and to see hatred, anger, and discrimination not so much as enemies but as *energies*—inside all of us—that can be embraced and transformed.

With the insights of interbeing and impermanence that we explored in the early pages of this book, we know that *a person is not only their views*. They are not "the side" they identify with. We may be surrounded by people whose political positions or views we disagree with; those people may even be in our workplace, community, or our own family. A view is only ever partial, not absolute. It is changeable, not permanent. And, if a view has come to be in someone's heart and mind, it's because something has "fed" that view. As Thay says, "Nothing survives without food." A view has been fed by algorithms and search results, by news feeds and clicks. The challenge is to train to see all views—including our own—as limited, impermanent, and open to change.

Confronted with a polarized society, the questions become: How can we have a mature dialogue across the divide? How can we have real communication? How can we be human, together? This has something to do with our capacity to listen, to cultivate openness, and to release our own views. Thay's teaching is strong here: in the light of interbeing, we cannot have the right without the left. In terms of views and dialogue, we should also have the perspective that our position has arisen in relation to their position. Our positions inter-are. "In true dialogue," he says, "both sides are willing to change." And that means, we have to be ready to release our views.

We might want to say to the other side, "You change first! If you're not going to change, I won't!" But, with the insight of interbeing, we know that our way of being, our openness, *already* changes the situation. The challenge is to be humble about the limits of our own perspective, to be open to learn something new, and to be genuinely curious about how others have arrived at the view they have. In the Plum Village tradition we train to respect the right of others to be different, and to choose what to believe and how to decide. And yet we are also committed to helping them transform fanaticism and narrowness through compassionate dialogue. We may need to be patient in order to understand. If we start from the position that "I am right" and "you are wrong," how will we ever reach a deeper understanding? We have to find a way to somehow turn down the reactivity and open up our hearts to see the human right in front of us and access our compassion and intention to understand their deepest fears, pain, and concerns. We remind ourselves to cherish the human even if we don't agree with their view. There is no "winning," no "solving" planetary problems without including everyone in the dialogue.

Catalyst for Change

Christiana Figueres, the architect of the historic 2015 Paris Climate Agreement, has shown that it is possible to put these teachings into practice and transform even the gravest of situations. Christiana is a student of Thay and brings his teachings on inter-being, deep listening, and personal reflection into the core of her work. Against the odds, her efforts in Paris were successful, and 195 countries came together to adopt the agreement. She says that Thay's teachings helped her "maintain an inordinate amount of calm in moments of total crisis in the negotiations." Without them, Christiana says she simply wouldn't have had "the inner stamina, the depth of optimism, the depth of commitment, the depth of the inspiration" to carry her through.

When I first saw Christiana in person, she was dancing in delight with a purple-robed bishop, on a stage in a church in Paris, on the eve of the conference. Different faith traditions and leaders had gathered in prayer and petition to underline the moral imperative for collective global action for climate justice. We had been invited to lead a meditation. At that point, no one knew if the conference would succeed. But we did know that world leaders were flying in that afternoon and maybe, just maybe, a miracle could happen.

It was while Christiana was dancing that I slowly walked up the packed side aisle, long-robed and shaven-headed, past the security guards, to wait for her. Eighteen months earlier, Thay had asked me to take care of offering her his support, encouragement, and love. He had recently suffered his stroke and couldn't be there, so a delegation of a dozen of us monastics had come to represent him. As Christiana stepped offstage, I joined my palms in greeting and opened my arms to embrace her. "This is a hug from Thay and from all of us. We're here for you." I cannot tell you how long the hug lasted, in the melee of dancers and press, with the security guards trying to escort her off to the airport, but I do know that there were tears and deep mindful breaths stretching out across space and time. When there is tough work to be done, we need all the spiritual strength we can get.

Christiana calls herself a stubborn, grounded optimist. For Christiana, optimism is not about anticipating a certain outcome, but about choosing *the kind of energy with which we enter the challenge* of the climate crisis. It's not about the result, and yet, as an energy, optimism will of itself change the result. There's a profound interbeing between ends and means.

As Christiana explains, "You enter into this engagement with optimism because you know that we *have* to do this. It's a sacred opportunity that *all* of us are having right now: to be alive and to be adults in the moment in which history and mankind are in such incredible transformation." While stewarding the Paris agreement, Christiana learned that "if you do not control the complex landscape of a challenge (and you rarely do), the most powerful thing you can do is to change *how you behave* in that landscape, using yourself as a catalyst for overall change."

Christiana describes listening deeply and sincerely as one of the most underrated yet transformational skills that she's brought to her climate engagement: "You cannot move to solutions without understanding the problem. Effective solutions will only be achieved if we honor and respect the differences of everyone; if we choose to understand the needs and pains of those across the table from us." She says:

I must say that there were many hard skills that we were engaged in, on the road to Paris, and for me the most powerful "soft skill" was the deep listening. We traveled to [speak to the leaders and climate negotiators of] almost every country in the world, mostly with questions—not to tell them what we thought they had to do—and with deep listening, to understand where they were coming from. It just opens a common ground that is not there if you're not listening.

It was one of the most powerful, personal growth experiences for me, because we're so used to thinking that another person's experience is only that person's experience. But if we do really engage in deep listening, we do very quickly realize that fundamentally we're all humans, and that the thoughts, the emotions, the fear, the anxiety, the grief that another person feels, is somewhere inside of us, also. Perhaps with a different colored coat, perhaps in a different language, perhaps in a different geography, but the sentiment is there, because it's a human sentiment. And so when someone is sharing out of their grief, if we feel very intentionally where that grief is in us, and begin to get in touch with that and heal that in ourselves, it just gives that expe-

rience with the other person a completely different quality, because you meet vulnerability with vulnerability; you hold hands with your mutual vulnerability. And once you have done that with another human being, the quality of that relationship moves to a completely different level, and from then on you can have technical discussions about megatons or whatever, but the deep root is there.

Radical Insight and the Arc of Love

Christiana also discovered that Thay's teaching on transcending the notions of "victim" and "perpetrator" had a powerful effect on helping unpack difficulties in the negotiations. In her own life, Christiana realized that, as for many of us, there were ways in which she *was* seeing herself a victim—in her case, in her difficult childhood and marriage. Reflecting on that in her spiritual practice, Christiana got the awakening that, "if I am labeling myself as a victim, I am immediately labeling somebody else a perpetrator." And very quickly that perpetrator turns back to you and calls *you* the perpetrator, and, before you know it, "you engage in this seesaw of victim-perpetrator and everybody is a victim *and* a perpetrator." In this dynamic, "you're a victim and a perpetrator at different points in time, with different people, with different situations." Christiana saw that happening in the negotiations. Developing countries "are objectively victims of climate change, but they don't have to stay there. We can get out of the victim-perpetrator dynamic." People were able to honor the reality of the historic responsibility, and at the same time embrace "a forward-looking common responsibility that has to do with the future of the planet and the future of all human beings on the planet." And, as Christiana began to bring herself out of her own dynamic of victim and perpetrator, she began to see the negotiations shift.

Compassion for all sides played its role. When asked whether she can summon love for the Koch brothers, wealthy industrialists who have profited from fossil fuels and opposed climate change legislation, Christiana is bold:

> *Here is the difficulty: we cannot be exceptionalist about it; we cannot. And while what they do makes me very angry, it does not dissipate my spiritual love for them because they are also on this planet. My challenge is to spread my arc of love over those that are close to me, those I love, those who are in my sphere. But also to spread that arc of love over those that I don't agree with, over the people that I have never met. I've had fascinating conversations with the Koch brothers, and there is something good, even in them. The moment that you engage in the blaming game and demonizing a person, a company, or a sector, you've lost your game. The moment you start doing that, you come down into a different level, from which it's going to be very difficult to emerge, because someone is going to have to win and someone is*

going to lose. And that's not the space that I want to work in.
I want to work in a space in which we all win.

Christiana maintains that all of us can contribute to a radical new future, in which everyone is heard, everyone is included, and everyone wins. This is the vision of Dr. King's "beloved community" in which even our enemies can be included. Christiana insists that each one of us has something to contribute:

The situation is not about power "over" anything; it's about power "for"—power for change, power for good. It's not about the privilege of "having" but about the privilege of being: being human beings, and the privilege of being in service of humanity. What better privilege do we all have? The privilege that we all share is the privilege of being alive right now and being humans at this incredible moment in time. By delving into our own vulnerability we humanize ourselves, and that is where we connect to each other and all of a sudden can discover where the power really lies—the power to change and the capacity to improve and work together are really there. Meeting each other and working hand in hand and walking hand in hand with who we are as humans, we walk so much faster and farther.

Bringing the Healing Home

There are parents who feel defeated, who feel that life has treated them unfairly, or who have been damaged in their childhood. Their hearts are full of frustration, injustice, and hatred. Because they don't know how to transform these violent energies within, they continue to make each other and their children suffer. The children bear the violence and don't dare to fight back. They accumulate this violence, and it gets unleashed whenever there is a chance. In countless families communication is blocked: no one knows how to listen, misunderstanding builds up, and everyone suffers. If we don't know how to practice to transform these energies within ourselves, our families, and our own generation, we'll destroy our future.

Our practice of deep listening and loving speech can help our father, mother, brother, sister, friend, lover, or partner. I have many young friends who, with the practice of mindfulness, have been able to embrace, heal, and transform their own difficulties and the difficulties of their parents. They give me a lot of faith that we still have a way out. We have a path and have nothing to fear. We need only walk this path together.

With a calm, sincere, and caring voice you can say something like, "Dad, Mom, I know that in the past few years, you've had a lot of difficulties, a lot of hurt that you haven't been able to express. And I haven't been able to help you. In-

217

stead, I've made things worse. I see that now, and I'm sorry. I promise from now on I'll stop blaming and provoking you. I just want to do something to help you suffer less. Dad, Mom, please tell me what it has been like for you. I want to understand what is difficult and what's weighing you down. I also have things I'd like to share with you. I know that in the past, I've been thoughtless, reactive, and unskillful. Please help me not to make the same mistakes again. There are so many things I want to realize, but haven't had the chance to do. I want to do the right thing, and I want you to be proud of me, but I need your help. Please tell me what I've done wrong and where I've been unskillful. I promise to listen, and I won't react like I used to. Please, help me." Speaking like this you are a bodhisattva Avalokiteshvara, ready to listen with compassion.

To succeed you must invest your whole heart in it. When your father *does* feel able to speak, he may not use loving speech because he has not yet learned how to. His words may be full of bitterness, anger, blame, or accusation, but keeping compassion alive in your heart, you stay protected. No matter what he says, don't get impatient or interrupt him or tell him he's wrong. It will only frustrate him and make him close up and start another argument. You remind yourself, "For now, I just listen. I will have time later to fill him in on what really happened so he can correct his wrong perceptions. Now is only for listening." If you can keep compassion alive in your heart during the whole time of listening, you will have succeeded.

If your father or mother is reluctant to speak, you can gently encourage them by saying something like, "Dad, I had no idea you have been having such a difficult time," or "Mom, I had no idea you went through so much."

Thanks to the practice of deep listening, thousands of parents have been able to reconcile with their children. To have someone listen to you for one hour, you feel much better, as if you've just taken a good dose of vitamins. In the future, when things have calmed down and the time is right, you can find ways to offer more information about what has happened so they can correct their wrong perceptions. It's best to do this slowly, offering only a bit of information at a time so that they can receive it and take time to reflect on it. If you get impatient, and want to set things straight in one sitting, they may not be able to handle it. Compassion goes together with patience.

Words That Heal

Once you have succeeded in listening to your parents, you can ask if they are willing to take a moment to listen to you. You can put into words things you've never had a chance to say. You have the right and the responsibility to speak about the deepest things in your heart, including your difficulties, hurts, or dreams. By coming back to your mindful breathing to embrace any strong emotions, you'll be able to use words that are skillful and easy to receive. The purpose is to help the other person correct their perceptions and to acknowledge your pain, difficulties, injustice, or dreams. And using loving speech helps the other person be receptive. We speak about our pain, our difficulties, and our dreams without blaming, accusing, or condemning, without bitterness or derision. We can ask for their support to listen without interrupting so that we have a chance to share everything that is in our heart.

I have faith that you can do this. When we're hurting, it's all too easy to utter words that damage ourselves and those we care about. But, with loving, skillful speech, we stop hurting each other and can begin to heal. Words don't cost anything but they can offer hope, strengthen our love, reestablish communication, and get us out of the depths of sorrow and despair. With just a few kind, loving words, you can offer happiness to so many people, including yourself. I have done this and I have succeeded. And many of my young friends have also succeeded. Don't think that only when you have money or power can you help others. You can help them right now, with loving speech.

You also have to speak to yourself with love. Many of us have suffered as a child, and those wounds have not yet healed. You can speak to yourself kindly and say, "I know you are there in me, my wounded little child. I'm sorry I've been so busy I haven't had the time to go back and take care of you. I am here now." With mindful breathing you keep your inner child company and help them heal. This is meditation. It's very urgent. We can say to them, "Look, we have grown up. We are no longer as vulnerable as we were. We can protect and defend ourselves very well." Tell the little boy, the little girl inside, "Don't be afraid. Let's go out and enjoy the sunshine, the beautiful hills and trees. There's no need to hide." This is the kind of meditation we can do to heal. There may still be a lot of fear and a tendency to withdraw. So, go home and talk to the little boy or little girl and invite them to enjoy the present moment with you. This is possible. With a few days of practice like this you can talk to the child inside and you can get the healing you need.

Learning the Art of Communicating

—T.D.

There is an art to embracing our strong emotions and anger so we don't cause more harm with our speech. It is possible to train ourselves to transform the energy of anger or rage into a fierce, loving compassion that can sustain our action and not burn us up. It's not always easy. Sometimes we may have to run our anger off with a jog, yell to the heavens, or fall to the ground in tears. That's okay. We live in difficult times, but, with a spiritual practice and collective support, we will find a loving way through. We can take comfort from the fact that Thay has written a whole book about anger; it's a feeling he's intimately familiar with. In his own life Thay demonstrates that it's possible to transform and redirect that energy into insightful, loving action.

As we have discovered in this chapter, deep, compassionate communication is not about a negotiation of needs or defending a certain position. With the radical insight of meditation, we realize that we are impermanent, and our views and positions are impermanent, and we know that the truth of our interbeing with others runs deep. It's important to be open and curious as we enter dialogue, ready to let go of and change our view. We cannot draw a hard line between ourselves and others or between our own transformation and the transformation of the situation. This is why true communication is possible, even in those very situations where it seems the most elusive.

Here, then, is the short text of the mindfulness training on loving speech and deep listening. You may like to read it slowly, reflecting on the ways it may be challenging or inspiring on your journey toward brave, compassionate dialogue.

The Mindfulness Training
on Loving Speech and Deep Listening

Aware of the suffering caused by unmindful speech and the inability to listen to others, I am committed to cultivating loving speech and compassionate listening in order to relieve suffering and to promote reconciliation and peace in myself and among other people, ethnic and religious groups, and nations. Knowing that words can create happiness or suffering, I am committed to speaking truthfully using words that inspire confidence, joy, and hope. When anger is manifesting in me, I am determined not to speak. I will practice mindful breathing and walking in order to recognize and to look deeply into my anger. I know that the roots of anger can be found in my wrong perceptions and lack of understanding of the suffering in myself and in the other person. I will speak and listen in a way that can help myself and the other person to transform suffering and see the way out of difficult situations. I am determined not to spread news that I do not know to be certain and not to utter words that can cause division or discord. I will practice Right Diligence to nourish my capacity for understanding, love, joy, and inclusiveness and gradually transform anger, violence, and fear that lie deep in my consciousness.

TRUE LOVE:
IS IT THE REAL THING?

Love Is Fuel

With the mind of love we have a heart on fire and the vitality and strength to do whatever we want to do. The mind of love is the energy of a bodhisattva who vows to become an instrument of peace, compassion, and well-being in the world. The mind of love can nourish and heal. The mind of love can help us protect the environment and the planet. The mind of love goes together with awakening, enlightenment. Understanding is the very foundation of love; it's another word for love because, once we understand, we already begin to love. There is a very deep connection between the heart and mind.

In Plum Village we have a very simple definition of love. We say that to love means to be there: to be there, first of all, for yourself, and for the wonders of life and the Earth all around you. And, once you are truly present, you can offer that presence to those you love. If you are not there, how can you love? If you can be truly present, that is already Zen. Meditation is to be truly present, to look deeply, and to recognize the people and wonders around us. And when we acknowledge the presence of the other person, they are happy and we are also happy. Sometimes, when I'm doing walking meditation on a full-moon night, I look up at the moon and smile and I like to say to the moon, "Thank you, moon, for being there. Thank you, stars, for being there." I acknowledge their presence.

If we look with the eyes of non-duality, we can establish a very close relationship between our heart and the heart of the Earth. When we can see that our beautiful Earth is not inert matter but a living being, right away something is born in us: some kind of connection, a kind of love. We admire, we love the Earth and we want to be connected. That is the meaning of love: to be one with. And when you love someone, you want to say, "I need you; I take refuge in you." It's a kind of prayer; yet it's not superstition. You love the Earth and the Earth loves you. You entrust your love to the Earth and you know it will never betray you. You would do anything for the well-being of the Earth and the Earth would do anything for your well-being. And this connection begins with mindfulness. You realize that you are here as a *child* of the Earth and you carry the Earth within you. Mother Earth is not *outside* of you; she is *inside*. Mother Earth is not your environment; you are *part* of Mother Earth. And that kind of insight of interbeing, of non-discrimination, helps you to be truly in communion with the Earth.

But some of us feel tired of the Earth and find it hard to love the Earth. We may resent, blame, or reproach the Earth for bringing us into a life of such suffering. We may wish we'd never been reborn or wish to be born somewhere else. But, with deep looking, it is possible to overcome all suffering and resentment and see the true nature of the Earth and of ourselves.

In the Belly of the Earth

The Earth is inside of us and we are already in the Earth. We don't need to wait until we die to return to the Earth. We need to learn how to take refuge in Mother Earth—it is the best way to heal and to nourish ourselves. We can do it if we know how to allow the Earth to *be*, within us and around us—just being aware that *we are the Earth*. And we don't have to do much. In fact, we don't have to do anything at all. It's like when we were in our mother's womb. We did not have to breathe, we did not have to eat, because our mother breathed for us and ate for us. We did not have to worry about anything.

You can do the same now when you sit. Allow Mother Earth to sit for you. When you breathe, allow the Earth to breathe for you, when you walk, allow the Earth to walk for you. Don't make any effort. Allow her to do it. She knows how to do it. Don't try to do anything. Don't try to fight in order to sit. Don't try to breathe in and breathe out. Don't even try to be peaceful. Allow the Earth to do everything for you. Allow the air to enter our lungs and to flow out of our lungs. We don't need to make any effort to breathe in or breathe out. Just allow nature, allow the Earth to breathe in and out for us. And we just sit there, enjoying the breathing in and the breathing out. There is the breathing but there is no "you" who is breathing in or breathing out. We don't need a "you" or an "I" in order to breathe in and out. The breathing in and the breathing out happen by themselves. Try it!

Allow yourself to be seated. Allow yourself to be yourself. Don't do anything. Just allow the sitting to take place. Don't strive in order to sit. And then relaxation will come. And you know something? When there is relaxation, then the healing begins to take place. There is no healing without relaxation. And relaxation means doing nothing, trying nothing. So, while "it" is breathing in—it's not "you" who is breathing in—you just enjoy the breathing and say silently to yourself "healing is taking place," and when "it" is breathing out, you say "healing is taking place." Allow your body to heal, to be renewed, to be nourished. This is what we call in Zen "the practice of non-practice."

If we know the practice of non-practice, we don't have to strive or fight. We simply allow our body to heal. We allow our mind to heal. Don't try anything. Allow yourself to relax, to release all the tension in your body and all the worries and the fear in your mind. Allow yourself to be held by the Earth, whether you are sitting, walking, lying down, or standing. Allow the Earth and sun to embrace you for healing to take place. Sit in such a way that you don't have to try to sit. You just enjoy your sitting deeply, nothing to do, nowhere to go. And, if you have half an hour sitting like that, you have half an hour of healing. Or, if you have one hour like that, you have one hour of healing. Or, if you have one day, that is one whole day of healing. It is possible. Make it pleasant, make it healing and nourishing. Don't try, don't make any effort. Simply allow yourself to take refuge in Mother Earth; she knows how to do it, and she'll do it for you.

Hungry for Love

All of us are hungry for peace, hungry for understanding, and hungry for love. We may have been wandering around, looking for someone who can give us love, but we haven't yet found anyone. However much we want to help society or the planet, we can't do anything if we haven't been able to fulfill our basic need for love.

In a relationship, in any kind of relationship, whether it is father and son, mother and daughter, or partner and partner, we all wish that the other person can provide us with these three things: some peace inside, some understanding, and some love. And if the other person can't provide us that, the relationship doesn't meet our needs, and we suffer. So, what should we do? A good question to ask ourselves is: How can *I* generate—how can *I* create—the energy of peace, understanding, and love?

Perhaps the person we love suffers. They have their own difficulties as well as their dreams and aspirations, but if we don't understand these things, we won't be able to offer them the second thing: understanding. How can I nourish and feed them with the right kind of love? We need to understand our own difficulties and suffering first, in order to be able to love others.

You want to love yourself, but do you have *the time* to love yourself, to take care of your body and your feelings? If you don't have time to do that, how can you help someone else? How can you love? Everything you do in your daily life can be an act of love. When you've been working on the computer for an hour, are you able to take time to stop working, go back to your body and enjoy breathing? This can be a powerful act of reconciliation, an act of love. Practicing mindful walking, or deep relaxation to release tension in the body, can be an act of love. It's possible to reconcile with yourself in a non-violent, gentle way. You say silently, "Dear body, I am here for you." You reconcile with your body, you reconcile with yourself. The question is: "Am I making time to take care of myself, to heal, so I can love, so I can serve, so I can help heal society?"

If we struggle to accept our actions and habit energies, we'll find it hard to love ourselves. There may be a feeling of hatred or anger toward ourselves. If so, in our meditation we can take time to look deeply into our actions, in order to recognize the seeds that led to them. The seed may come from your ancestors. The energy of your father, grandfather, or great-grandfather may have been taking that action with you; or maybe it was your mother, grandmother, or great-grandmother. It's important to remember that you are made of non-you elements. There may be seeds of actions that were planted in your lifetime, and other seeds that were planted long before. So you have to look into all the actions you have taken—whether they were good, bad, or neutral—and see them in the light of no-self.

Sometimes a habit energy pushes us to do or say something. We don't want to do it, but we can't help it. Perhaps we didn't even realize we were doing it at the time. It is as though the habit, the seed, is stronger than us. When you practice mindfulness, you have a chance to introduce the element of awareness. It's very interesting. If you're curious enough, you'll be able to look deeply, be concentrated, and see the roots of that action. And, if you recognize that the action isn't beneficial for yourself or the world, you resolve not to repeat it. In this way we're really practicing for all our ancestors and future generations, and not just for ourselves. We are practicing for the whole world.

Love Without Boundaries

Love's true nature is inclusiveness, non-discrimination. If there's still discrimination in it, it's not yet true love. In Buddhism we speak of "love without boundaries"—the Four Immeasurable Minds.

The first is *maitrī*, which can be translated as "loving kindness, friendship, or companionship." Maitrī has the power to offer happiness. If love doesn't bring happiness, it's not true love. And so maitrī is not just the *willingness* to offer happiness but the *capacity* to offer it. If love can bring happiness, it's true love. You need to look to see whether maitrī is there in your relationship. It may be that maitrī is there but it's still a bit weak. If it's still weak, you have to help it grow. Love is something to be cultivated.

The second element of true love is *karuṇā*, compassion. Karuṇā is the capacity to bring relief, to remove suffering. We have to look to see if our relationship has karuṇā or not: the capacity to alleviate and transform the pain and suffering. If our friendship, our love, our relationship has karuṇā, that means it is true love. But if our love doesn't alleviate our pain but makes it worse, then it's not true love.

Compassion has the power to heal, and there is never enough of it. Some psychotherapists talk about "compassion fatigue." But, if people run out of compassion, it's because they don't know how to keep producing it. The fatigue doesn't come from having too much compassion but from running out of compassion. Compassion is a kind of power, a kind of energy we need to keep generating every day. There's a way of helping others that helps ourselves at the same time, and we should learn it. It's just like the tree in the front yard: everything the tree does to be and remain a healthy, fresh tree is for the good of the whole world. In the same way, everything you do to keep your own compassion alive is also good for the other.

Creating a collective energy of compassion is one of the best things we can offer to humankind and other species. We should learn how to do it. Compassion is made *only* of non-compassion elements, and so the art is to make use of those non-compassion elements like fear, anger, and despair to *create* compassion. If we know how to handle the suffering we have in the world, we can transform it back into compassion and love.

The third element of true love is *muditā*, joy. True love always brings joy, to ourselves and to the other person. If our love makes us cry every day, it's not true love. Love should bring joy. We can ask our loved one about this. Is our love bringing joy? To love means *to be there* for the one you love. We don't need to buy anything to bring them joy; we simply need to offer our full presence.

The final element of true love is *upekṣā*, inclusivity. We no longer exclude; we include everyone. Our love benefits everyone, not just one person. You and the person you love are one. Your suffering is their suffering; their happiness is your happiness. You can't say, "That's *your* problem." There's no individual happiness; there's no individual suffering. This is the meaning of no-self.

One day the Buddha was holding a bowl of water in his left hand and he was holding a handful of salt in the right hand. He poured the salt into the water and stirred. He asked the monks, "My dear friends, do you think that you can drink that water? It's too salty! But, if you were to throw that amount of salt into a big river, it wouldn't make the river salty at all, and thousands of people could continue to drink the water in the river."

Someone who has a great heart, a big heart, a lot of compassion, doesn't suffer anymore. The things that make others suffer don't make such a person suffer, in the same way that a handful of salt can make a bowl of water salty but can't make the river salty at all. When our heart feels small, the practice is to grow our compassion and inclusiveness by taking care of our own suffering first.

I would like to help the Buddha, and add two more elements of true love: *trust* and *reverence*. Of course, these two elements can be found in the four, but to make it more obvious, we have to mention their names. When you love someone, there has to be trust, there has to be confidence. Love without trust is not yet love.

And one of the things you trust is that you have buddha-hood, you have awakening in you. You trust that you contain the whole cosmos; you are made of stars. And that is why you respect yourself and offer reverence to yourself. And, when you look at another person, you see that they are also made of stars. They are a wonderful manifestation. They don't appear only for a hundred years: they carry eternity within them.

True love is something very real: we can recognize whether it is there or not. We need time and we need some practice to cultivate it. True love is there *as a seed* in every person, but we need to water the seed to help it grow. When we're able to cultivate all these elements—loving kindness, compassion, joy, inclusivity, trust, and reverence—love will spring up in our body and mind. We become filled with love.

Love is like light. And, just like a light bulb, as soon as there is electricity, the light already radiates. Love shines. Love illuminates without discrimination. This is the true love taught by the Buddha: the kind of love that doesn't make us suffer any more, the kind of love that brings relief, nourishes, and heals. You start by loving one person, but if it's true love, it will grow to encompass everyone and all beings, not only people but all animals, plants, and species.

There's an Art to Being a Soul Mate and They Don't Teach It at School

—T.D.

The monk Ananda once asked the Buddha, "Is it true that having good spiritual friends is about half the spiritual path?" We know meditating is important, we know right action is important, but how important is it to have good friends? Fifty percent? The Buddha's answer was "No, Ananda. Having good spiritual friends isn't half the path: it's the whole path." Thay's way of expressing this is to say, "Nothing is more important than brotherhood and sisterhood." Perhaps our generation might say, "Nothing is more important than love, friendship, and solidarity."

When I first went to Plum Village in my early twenties, I heard Thay teach about the power of aspiration: the importance of making a vow and commitment to live in a certain way. We identify our ideals, place them in front of us, intentionally, and commit to live up to them. The energy of commitment gives us determination to transform the habits that hold us back. And Thay also spoke about the importance of commitment in relationships: how we need to commit to one another and invest in one another; we can't just be fair-weather friends. We have to promise to be there for each other in the most difficult moments—that is genuine solidarity, real friendship, true love.

At that time, I was in a relationship, and I couldn't figure out how the teachings on impermanence went together with the teachings on love: How can you commit to love someone if you're impermanent and they're impermanent? Surely the commitment itself is impermanent as well? During a twenty-one-day retreat, my partner and I asked Thay this question: "How can we make a commitment to each other, in the light of impermanence?" Thay smiled. He looked at each of us in turn and asked, "Are you the same person you were yesterday? The truth is that you're neither the same nor different; you're not the same, but you're not an *entirely* different person either."

We think we know the person we love, but perhaps we shouldn't be so sure. Each of us is an ever-changing stream: from one moment to the next, our body is changing, our feelings are changing, and our perceptions are changing. It's scary. We don't necessarily *want* our loved ones to change. Perhaps there are a couple of very *specific* things we'd like them to change, but in general, we're afraid of change: we're afraid that we'll lose the person we love or the person who loves us back.

Thay taught us to cultivate the kind of insight in which everything *is* changing, and the challenge is to always help it change for the better not the worse. He taught us to see each other as a garden. There may be different kinds of plants, flowers, and trees in our garden, and, as the gardener of each other's garden, we can help create conditions for sunshine, rain, and shade, as needed. And, as a gardener, it's also our responsibility to take care of the weeds and our beloved's compost. It's our job to help *transform* their compost and not be afraid of it, to make good use of it to nourish the garden and make it more beautiful. "Love is organic," Thay said, with a grin.

After receiving this answer, we learned above all to not be afraid of each other's compost and to not try to hide it. We became "compost comrades." Practicing deep listening and loving speech, we discovered that helping each other transform the junk of our habits and flaws could be a joyous and messy shared enterprise. We also learned that even the original commitment to each other is organic and alive, and needs to be nurtured, so it can grow, evolve, and continue to nourish both the relationship and our shared aspirations. The commitment itself also needs food, freedom, and space to grow.

I remember how shocked I was when Thay gave a talk at the House of Lords in London and he opened with a line about love. We had invited politicians and journalists to come to "a talk on mindfulness and ethics." Thay fearlessly cut to the chase, to the heart of the matter, saying, "We all know that love is a wonderful thing. When love is born in our heart we suffer less right away and we begin to heal."

That day I learned that, for Thay, the kind of compassion we hope to bring to our engaged action and service is no different from the love of our intimate relationships. Compassion isn't a professional skill. It's not something we can instrumentalize, saying to ourselves, "Ah, if I'm compassionate, that will be the most effective way to get what I want." True love is much bigger than that: it's an energy that changes the one who is loving and the one who is loved. True love is generous, tolerant, and forgiving.

The Buddha said that if we don't have peace in our relationships, there's no way we'll be able to have peace on our cushion. And, in order to sit with ourselves in peace, we need a certain softness, acceptance, and kindness toward our own body and mind that we can then extend to those around us, and to our planet. This is the "arc of love" that Christiana Figueres mentions; we want to spread that arc wide. But, sometimes, the hardest place to reach is our own heart, our own loneliness, and our own self-criticism. How can we help let the light and love in?

Love Meditation

Love meditation in the Buddhist tradition should be directed to yourself first. The practice of self-love recommended by the Buddha is simple, effective, and not difficult to practice. It begins with focusing our attention on what we really want. We should be *aware* of what we really want the most. The Buddha proposed we contemplate:

May I be peaceful, happy, and light in body and in mind.

How can happiness be possible if we are not peaceful and light in our body and mind? If it feels too heavy in our body and our mind, if we don't feel peaceful, how can we be happy? I *want* to be peaceful. I *want* to be light in body and mind. If you know what you really want, you can offer it to yourself.

The next contemplation is:

May I be safe and free from accidents.

There's so much violence and so many accidents in the world. We say to ourselves: I want to be protected, I want to be safe. And if I know I want these things, my practice of mindfulness can help me bring peace and lightness into my body and mind. Mindfulness can protect us.

And then we contemplate:

May I be free from anger.

When I am angry I'm not happy. I want to be *free* from anger. And the practice can help. When anger takes hold of me I feel burned. I want to be free from unwholesome states of mind, including anger, despair, jealousy, fear, and worries.

Next, we contemplate:

May I know how to look at myself with the eyes of understanding and love.

Sometimes we cannot accept ourselves, we hate ourselves, we're angry at ourselves. We're not satisfied with ourselves. We can't look at ourselves with the eyes of compassion. To look at others with compassion, you first need to be able to look at yourself with compassion, and accept yourself as you are. We practice not to blame ourselves, but to look deeply at the roots of our suffering, and all the causes and conditions that have led us to suffer, so we can accept ourselves with compassion. Once you are able to accept yourself, you suffer less right away. We are learning to love ourselves, to take care of ourselves.

The next step is to contemplate:

May I be able to recognize and touch the seeds of joy and happiness in me.

There are seeds of happiness and joy in us and when they are watered they give rise to the energy of joy and happiness. The contemplation here is to recognize these seeds. This is a way of loving ourselves. Our friends can help touch and water the seeds of happiness and joy in us, but we can do it for ourselves, too. We actively recognize these positive seeds, and we know how to breathe and walk in such a way that helps them manifest.

Next, we contemplate:

May I learn how to nourish myself with joy each day.
May I be able to live fresh, solid, and free.

We all need joy. We all need happiness. We want to be solid. We know that solidity is the ground of happiness. If we're too unstable or fragile, happiness will not be possible. So we resolve to cultivate solidity, freshness, and freedom. There are many meditations to help cultivate our solidity, freedom, and calm.

Next, we contemplate:

May I not fall into a state of indifference.

We don't want to be someone who doesn't care and who is indifferent. We want to be concerned for our own well-being and others' well-being. Although we don't want to be indifferent, we also don't want to be caught in either of the two extremes of attachment and aversion. When we're caught in an infatuation, a craving, or an addiction, we suffer. And when we're angry about something, we suffer. Both craving and aversion rob us of our freedom and happiness.

So this meditation is to contemplate those things we truly want for ourselves. And the Buddha teaches us that we can offer them to ourselves right away. This is the first step of love meditation.

May I be peaceful and light in my body and in my mind.
May I be safe and free from accidents.
May I be free from anger, unwholesome states of mind,
fear, and worries.
May I know how to look at myself with the eyes of
understanding and compassion.
May I be able to recognize and touch the seeds of
joy and happiness in me.
May I learn how to nourish myself with joy each day.
May I be able to live fresh, solid, and free.
May I not fall into a state of indifference or be caught in
the extremes of attachment and aversion.

At first, we train to offer love to ourselves. And then, after a few days of practice, we take the next step and practice love toward another person. You have offered it to yourself already, and now you can offer it to them. We contemplate:

May they be peaceful and light in their body and mind.
May they be safe and free from accidents.
May they be free from anger and unwholesome states of
mind like fear and worries.
May they know how to look at themselves with the eyes of
understanding and love.

You give rise to the wish to help them do that. So this is the second step of the practice: love directed to another person. And the third step is to address our love to everyone, to all beings, not just one or two people. True love is without boundaries; it is a limitless mind. You open up to include everyone:

May all beings be peaceful and light in their body and mind.
May all beings be safe and free from accidents.

Keep Your Loneliness Warm

You feel lonely because you have not seen the connection between yourself and other beings. You have not seen the connection between you and the air, the sunshine, the water, the people, the animals, the plants, and the minerals. You are lonely because you believe there is a separate self. The insight of interbeing can help solve the problem of loneliness.

Everything is there for you. That is the truth. The sunshine is there for you. If there is no sunshine, there is no life on

Earth, and you couldn't be here. So, you have to see the deep connection between you and the sunshine. You are made of sunshine. Is the sunshine lonely? The water, the air, Mother Earth, the stars, the moon—they are all there for you. You can train yourself to breathe, walk, and sit in such a way that you can get connected with the stars, the trees, the air, the sunshine.

Life is a wonder, and your body, your feelings, are also wonders. And, if you know how to connect yourself with all of these things, you will not be lonely. The sunshine has the power to love. And we human beings also have the power to love. If the sunshine loves us, then we should be able to love the sunshine back. If the trees love us, then we should learn how to be able to love the trees. And, if we know how to love, we don't feel lonely anymore.

To feel sadness, loneliness, is not something bad. All of us feel sad or lonely from time to time, and we can learn to come back to ourselves to embrace our loneliness. It is a wonderful practice. Sometimes you can feel very comfortable embracing your loneliness, keeping your loneliness warm. You don't have to push your loneliness away. Your loneliness is there, and you *accept* it. You breathe in and out to be truly there and you embrace your loneliness. Sometimes we want to be alone and hold our loneliness. We feel we can be there for ourselves, and we don't need anyone else to help us. We have the capacity to take care of ourselves.

The teaching on true love is very clear. To love means to

be there and listen deeply to see the suffering and loneliness of the other person. Once we feel there is *one* person who can understand us, our loneliness disappears. You are lucky if there is someone who can truly understand you, your suffering, difficulties, and loneliness. You are receiving a gift from them, and that gift is the power of understanding. And you have to offer the same gift in return. You can ask, "Do I understand you enough? Please help me understand you." Love is a gift that can make the other person not lonely anymore.

Three Kinds of Intimacy

If you feel lonely and cut off, if you suffer and need healing, you cannot expect the loneliness to be healed and comforted by having sexual relations. It will only create more suffering for both you and the other person. First, you need to learn how to heal *yourself*, to be comfortable within and cultivate a true home inside. And, with that true home, you'll have something to offer to the other person. And *they* need to do the same: to heal themselves, so they can feel better and at ease and share that home with you. Otherwise, all they'll have to share is loneliness and ill-ease, which cannot help heal you at all.

There are three kinds of intimacy. The first is physical and sexual, the second is emotional, and the third is spiritual. Sexual intimacy cannot be separated from emotional intimacy;

they go together. And, if spiritual intimacy is there, the physical, sexual intimacy will be meaningful, healthy, and healing. Otherwise, it will only be destructive. We're all seeking emotional intimacy: we want to be in harmony, to have real communication, mutual understanding, and communion.

Accepting your body is a very important practice. You are already beautiful as you are; you don't have to be anyone else. When you can accept your body and make peace with your body, you have a chance to see your body as your home. You have peace, you have warmth, and you have joy. Building a true home within feels wonderful.

Sensual pleasure and sexual desire are not love, and yet our society is organized in such a way as to make sensual pleasure the most important thing. Companies use your craving to sell products and make profits. But sexual craving can destroy body and mind. What we all need most is understanding: mutual understanding, trust, love, emotional intimacy, and spiritual intimacy. Sexual intimacy *can* be a beautiful thing—if there is mindfulness, concentration, insight, mutual understanding, and love. With communion and mutual understanding on the emotional and spiritual levels, physical, sexual intimacy can become sacred.

Love is very intimate. There are very deep zones in our spirit. There are sacred places in our being. There are things we don't want to share with just anyone, things we want to keep secret to ourselves—very deep feelings, very deep, sacred memories. We want to guard these things closely. Only when we have found someone who has the capacity to understand us deeply do we feel able to open our heart. We invite the other person to enter and to visit these sacred zones. It's a communion, a very deep communication. And this can only happen with true love. We let them enter into our world, and we are ready to share with them everything we hold sacred. We reserve consent for only those who understand us.

We should not think that the body is one thing and the mind is a completely different, other thing. Mind and body inter-are. You cannot take the mind out of the body or the body out of the mind. Even modern medicine is beginning to operate on this insight. When there is deep communion, deep communication, deep understanding between two people, the coming together of the two bodies will only *enhance* that communion. There is the awareness that the other person is very precious, not only as a body but as a mind. And you have a lot of respect for them, both their mind and their body, and the body is not an object of pleasure. True love should always include a sense of reverence and respect.

Are They "the One"?

In a relationship, if the other person is not able to listen to you and understand you, then you know that person is going to make you suffer in the future. This is very clear, very simple. If, talking with them, you find they're not able to listen, they always interrupt you, trying only to make their point, and they're not interested in understanding your suffering and difficulties, then you know that person won't be able to understand you or bring you happiness in the future. They may be attractive, they may have status, but if they're not able to listen and understand you, they are the wrong person. They'll only give you a hard time.

This is very easy to detect. You only need fifteen minutes to find out whether they have the capacity to listen and understand you. And you should ask yourself the same questions: Do I have the capacity to listen to them? Am I interested in understanding their suffering? And if you see that you have that desire to understand their suffering and the desire to help them suffer less, then you know you can go on with the relationship because you are equipped with that good intention.

It's very concrete. In a relationship, we can already know whether we'll be happy in the future or whether we'll suffer a lot, just by reflecting on our *own* capacity and intention to listen and understand, and the other person's capacity and intention to listen and understand. This is very important. It's fundamental.

Misunderstanding is a daily problem for everyone. Not only does the other person misunderstand us but we misunderstand ourselves. We don't know who we are, so how can we expect the other person to know? If we don't take time to observe ourselves, we won't understand who we are, and we won't be able to see our strengths and weaknesses, and we'll have a wrong perception of ourselves. And yet we want the other person to have a good, correct perception of us! This is difficult. We can say, "I know I still haven't understood myself completely, so, if you've seen something, please let me know. Please help me understand myself better, so I can then also understand you." That is an attitude of openness. With mutual understanding and good communication, happiness is possible and the relationship can last.

It's very important to discover our partner's deepest desires and aspirations. If you don't understand their deepest desire and motivation, if you don't feel you can support them in realizing that intention, you won't be able to be a true friend. And in return you have to tell them about *your* deepest motivation to do something beautiful, meaningful with your life, to see whether they are ready to support you.

There should be a perfect harmony in the way you live and the way you love. You have to be able to talk with your beloved about the way you earn your living, your worries, your concerns about society or the planet. That is your duty if you are a true lover. It's not enough just to bring home your monthly salary; you have to bring home happiness and peace and help each other cultivate that happiness and peace. That is why there should be a continuous dialogue for your love to grow.

Learning the Art of True Love

—T.D.

It's unusual for a Zen master to talk about intimacy, but Thay teaches about it because a lot of people ask. In the monastery, part of our monastic training is learning to stay creative and healthy as we balance the three different kinds of energy: sexual, breath, and spirit. There is a lot to discover and understand about body and mind.

One of Thay's most remarkable contributions to the meditation tradition is his emphasis on the importance of nurturing our life force and vitality—bringing a "spring warmth" to our practice. Meditation and mindfulness are not intended to make us like dead wood; the practice is there to help us feel more alive and to help us make good use of our energy of vitality and love as a force for good in the world.

In Plum Village every summer we host a weeklong mindful living retreat for hundreds of young people from all over the world. We meditate together and train in mindful breathing and relaxation. We go for hikes, make music, enjoy bonfires, and work together on our organic farm. One of the most powerful sessions is when we create safe spaces to talk in smaller groups about experiences of sexual intimacy and the pressures and vulnerabilities we have all felt in sexual relations. Discussions often become very heated. There are those who insist that sex can and must be separated from love, and others who testify to the emotional pain of moments when their

251

bodies were exploited for pleasure. There are those who can't imagine a healthy life without porn, and others whose relationships have been destroyed by it. Is it only a question of communication and consent, or is it something more? Everyone has different views, perspectives, and lived experiences, and it's important to honor that. It's also important to be brave enough to acknowledge when we have been hurt and when others have been too. Listening deeply and looking deeply can help us develop the qualities of presence and communication we need to navigate a healthy relationship with our own hearts, bodies, and loved ones.

Here is the text of the mindfulness training on true love. Even if you find the words challenging, they are an invitation to reflect on the pain points in our love and relationships, and to create conditions for healing and fulfillment. As with the other four of the Five Mindfulness Trainings, it's not intended as a hard-and-fast rule but as a contemplation to help us look deeply and grow.

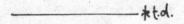

The Mindfulness Training
on True Love

Aware of the suffering caused by sexual misconduct, I am committed to cultivating responsibility and learning ways to protect the safety and integrity of individuals, couples, families, and society. Knowing that sexual desire is not love and that sexual activity motivated by craving always harms myself as well as others, I am determined not to engage in sexual relations without true love and a deep, long-term commitment made known to my family and friends. I will do everything in my power to protect children from sexual abuse and to prevent couples and families from being broken by sexual misconduct. Seeing that body and mind are one, I am committed to learning appropriate ways to take care of my sexual energy and cultivating loving kindness, compassion, joy, and inclusiveness—the four basic elements of true love—for my greater happiness and the greater happiness of others. Practicing true love, we know that we will continue beautifully into the future.

North Star

Taken together, the Five Mindfulness Trainings are a Buddhist contribution toward a new global ethic. They show a way to bring about collective awakening, so we can live together and act together in such a way that can save the planet and make a future possible for generations to come. We urgently need to learn to change our way of living so there is more mindfulness, more peace, and more love, and each one of us can do that beginning now, today.

If you're inspired to be a bodhisattva of our time, protecting and safeguarding what is beautiful, with these Five Mindfulness Trainings in your heart, you will have energy and insight you need to be a bodhisattva on the path of action. Whatever culture you are from—and whatever your spiritual roots—these trainings can be the foundation of your life and represent your ideal of service. Their nature is non-sectarian and universal.

The Five Mindfulness Trainings are a practice of real love. We want our heart to continue to grow so that it embraces not only one person but the whole world. That is the love of an awakened person: love without boundaries. And it is possible. If you follow the path of true love, very soon you can include many others and realize a great aspiration.

I have learned that my home, my country, is the whole planet Earth. I do not limit my love to a tiny piece of land in

Asia: Vietnam. And I have experienced a lot of transformation and healing because of that vision. Your love may still be too small. You have to enlarge your heart and allow your love to embrace the whole planet. That is the love of buddhas and bodhisattvas and great human beings like Mahatma Gandhi, Martin Luther King Jr., and Mother Teresa.

You don't need to be perfect. What's important is that you have a path to follow, a path of love. If we get lost in a forest and we don't have a compass at night, we can look at the North Star in order to go north, to get out. Your purpose is to get out of the forest, it's not to arrive at the North Star. So, having a path, a direction to go in, is what we need most. And then we have nothing more to fear.

To open the way for the next generation, we should follow a new path of solidarity and compassion, of brotherhood and sisterhood. We have to come together to do this. We have to take the situation into our own hands. And don't wait for the government: you'll have to wait a very long time.

wake
up
together

PART 3

COMMUNITIES OF
RESISTANCE
A NEW WAY
OF BEING
TOGETHER

A Place of Refuge

After his enlightenment, the first thing the Buddha did was to look for his friends and set up the first group of practitioners, the first sangha. We should do the same. We should create an island of peace and togetherness wherever we live. A sangha is a refuge. It is an island of peace. It is a community of resistance against violence, hate, and despair. We all need a place like that to go to.

The word "sangha" simply means community. A political party is a kind of sangha, a family is a sangha, a corporation, clinic, or school can be a sangha. When we use the word "sangha," we mean a community where there is harmony; where there is mindfulness, concentration, and insight; and where there is togetherness and joy. There's no struggle for power and no division. Any such community can be a sangha. And, if you have a dream, no matter how determined you are, you will need a sangha, a community, in order to realize it.

The national motto of France is *Liberté, égalité, fraternité*: freedom, equality, fraternity. In my own life I have seen that, without fraternity, without brotherhood and sisterhood, without solidarity, you cannot do anything. With a community we can get the strength we need to create equal opportunity for all and inner freedom—freedom from despair and from the prisons of the past and the future.

Many years ago, I spoke to my friend Father Daniel Berrigan (a Jesuit priest, poet, and peace activist) about communities of resistance. We are constantly being invaded by negative things in society. We're assaulted day and night by what we see and hear, and the negative seeds in us continue to grow. That is why we need to reflect on how to organize communities to protect us. Creating a healthy, compassionate environment for each other is very important.

You have your bodhicitta: the mind of love, the will to transform, the desire to serve. You have woken up and you realize you want to live differently. Bodhicitta is like rocket fuel: it's so powerful it could send a rocket to the moon. But, in order to help the energy of bodhicitta be strong and sustainable, we all need a sangha, a place of refuge. We need a community to support us in our practice. A community can offer the kind of environment we need to nourish and strengthen our aspiration so that we can get out of our personal and collective situation.

To succeed on our path, we have to take refuge in a community. It's not a matter of devotion but of action that can help us go in the direction of healing. And we have to stick to this community, entrust our whole being to our community, and allow our friends to carry us, like a boat. A community, a sangha, is a boat, and everyone in this boat is practicing and going in the same direction. We are part of the boat and we allow the boat to carry us. Without the boat we would sink. This is my experience. With a sangha you will never feel alone, you will never feel lost.

As soon as we wake up and we give rise to the aspiration to live differently, our healing and transformation begins right away, but, if we want to *continue* to heal, we need an environment that is conducive to healing. We are a warrior on our path, yet we still need a community to continue to be a warrior.

As soon as we have found our path and a community, we have peace already. Simply to be on the path, we can have peace, and steadily that peace will develop and grow. It's as though we've already boarded the train; there's no need to run anymore—all we need to do is to sit there and wait for the train to take us to our destination. Entrusting ourselves to our community, allowing our friends to carry us, we feel at peace.

Some Rice, a Few Clothes, and Good Spiritual Friends

When we founded the School of Youth for Social Service during the war in Vietnam, hundreds of young people enrolled to live, practice, and serve together. Each one of us had our dream for Vietnam, and we were living out our dream every day. Even though the living conditions were very simple, even though we had no salaries, or private homes, or cars, we were able to build true friendship and solidarity. In the midst of the bombs and violence, we established pilot villages and a movement for rural reconstruction, raising the standard of living by contributing to the economy, infrastructure, education, and healthcare. With that kind of brotherhood and sisterhood, and with a shared dream to realize every day, there was no need to chase after wealth, fame, power, or sex. Working in a situation of war was difficult and dangerous. Even though we faced immense challenges, we never gave up. We trained our social workers in mindfulness, and I wrote *The Miracle of Mindfulness* as a manual to help them stay healthy, focused, and compassionate, to nourish their aspiration so they could have enough joy and peace to continue their work.

In communities of mindful living we're able to demonstrate that it is possible to live happily and with few possessions. A few hundred people can live together and devote their time and energy to building solidarity, compassion, and love. And they can organize for others to visit and experience a kind of happiness and joy that comes from living deeply rather than from consuming.

We should try to create these kinds of communities in the countryside. It doesn't need to be Buddhist. You come together and live in such a way that can protect the Earth and protect the environment. You can share houses and equipment, establish your own school, and cultivate the land. Establishing small communities like that we can truly generate the energy of brotherhood and sisterhood—the kind of energy that can't be bought in the supermarket.

Master Linji taught that you don't need much to find freedom. He said you only need "a bowl of brown rice, some clothes on your back, and with these, you set out to invest all your energy in finding good spiritual friends." Don't look for anything more. That's all you need. Good spiritual friends are the kind of people who can help us open our eyes and be our true selves.

Six Principles of Togetherness

—T.D.

Thay taught us that the strength of our community depends on how much harmony we have. If there's no harmony, we lose a lot of energy pulling each other in different directions, and we'll have no energy left to realize our shared aspiration. In Plum Village, Thay emphasized the importance of cultivating the Six Harmonies: *lục hòa*, 六和. These principles from the Buddhist tradition he sometimes translated as the "Six Togethernesses." They represent six areas of community life and collaboration where we can actively develop harmony.

1. Physical Presence

It's important to show up for each other. It's important to invest our time, energy, and physical presence in being there for one another and for our shared aspiration. We want to be someone our friends and colleagues can count on and take refuge in. In the Buddhist tradition, this principle is sometimes described as the harmony of living or gathering "under one roof." It's the collective strength that arises when we commit to come together, to physically show up on screen or in real life, and to invest ourselves in developing collective energy and insight. We can ask ourselves, in relation to our own community or network: Am I showing up enough? Am I putting my heart into this? Am I someone others can count on? How can I create conditions to make it more inspiring and nourishing to spend time together?

2. Sharing Material Resources

The more we can share, the more we can be in harmony. It may be as simple as everyone pitching in for snacks, utilities, or expenses, or it may be as much as shared spaces and investment. In Plum Village, all our resources are held in common, and everyone contributes to decisions about how those resources are spent. It's very bonding. It's a concrete way to practice interbeing. It helps us let go of the idea that something belongs to us as an individual and helps us make decisions for the collective benefit. We can ask ourselves: Am I sharing enough? Is there anything I'm being possessive about, that is getting in the way? Is there something more we could share to reflect our trust and commitment to each other?

3. Sharing Ethical Principles

Whether it's a simple mission statement, a concrete commitment to non-violence and inclusiveness, or a specific code of conduct to establish some red lines and ways to resolve disputes, it's essential to agree on the values and direction that lie at the heart of our being and acting together. They function as a compass to guide us, a container to hold us. The Five Mindfulness Trainings are a powerful blueprint, used by thousands of communities around the world as a North Star. Depending on what unites and holds your community together, you may be inspired to develop your own version of them for your own context, culture, or faith.

4. Sharing Insights and Views

Thay always teaches us that being tolerant, inclusive, and open to diverse views is an essential principle to avoid dogmatism, discrimination, hatred, and violence. To share insights and views, here, doesn't mean that we necessarily hold the same ones; it means we are committed to creating an environment in which all views and voices are safe to be expressed and heard. We do our best not to impose our views on others. We try to create space for a diversity of views and be open to seeing things in a new way. We need to be ready to let go of what we already know in order to be open to others' insights and experiences. In this way, an authentic collective insight and "harmony of views" can naturally arise.

5. Sharing from the Heart

There's one word for both heart and mind in Vietnamese and Chinese, and this principle is therefore sometimes also called "harmony of thought." It means that we practice expressing our own experiences and truth deeply and honestly, and do our best to create space for others to speak from the heart too. It's a profound way to build trust and solidarity. What's really going on for me and for you? What's our deepest concern for our community? What are our deepest dreams? When we can share our insights and views from a place of sincerity, grounded in our own experience (and even our fears), it's much easier for friends and colleagues to hear, take it on board, to build a harmonious understanding.

6. Compassionate Communication

It's important to make a commitment to each other to guard our speaking, to practice restraint so we don't create harm. Means and ends go together; we don't just "speak the truth" (which is only our *perception* of the truth) without responsibility for the consequences. Bald, direct, unskillful, so-called truths may be violent in their effect, and they can damage trust. In Plum Village we train to speak out, with calm and compassion, to express our view, and then we train to let go. We do our best not to fight for it. If, in a meeting, a strong emotion comes up, we step outside for a ten-minute walk before coming back to express ourselves with more calm. When communication between two or more of us has, for whatever reason, become blocked, we do our best to set up a separate session of deep listening in order to understand the root of the friction, each other's experience of the situation, and our deepest concerns.

In our Plum Village residential communities we come from many different nationalities, backgrounds, and cultures. And yet everyone has "a seat at the table," or a cushion in the circle. All of us have chosen this path of training and practice, and that is the ground of our togetherness. Thay taught us to see every person we encounter as "a country to discover." Everyone has their value; everyone has a talent to be revealed and cultivated. This is true in any community, any team of collaborators. The challenge is to create conditions for each flower in the garden to flourish in their own unique way.

Over the last four decades, Thay has built a grassroots network of communities of mindful living, with thousands of local meditation sanghas in the Plum Village tradition, supported by a dozen monasteries in the US, Europe, and Asia. The "Earth Holder Sangha" is the branch of the Plum Village tree that advances Thay's teachings on engaged Buddhism, racial and social justice, and interbeing with Mother Earth. Earth Holder sanghas meet every month, online or in person, to practice meditation, to share insights and experiences on applying Earth-loving practices in daily life, and to advance Earth-healing compassionate direct action.

The "Wake Up Movement" is an international network of young people committed to creating local "communities of resistance" guided by the Five Mindfulness Trainings. They meet weekly or monthly to practice meditation and mindfulness, to create a refuge of togetherness and inspiration, and to bring healing and compassion to our fractured society. The "ARISE Sangha," another affinity group in the Plum Village Tradition, explores the dynamics of race, intersectionality, and social equity as Dharma doors to collective awakening.

Building strong communities for healing, awakening, and Earth justice requires including diverse experiences and perspectives. Thay's teachings on togetherness, harmony, and inclusiveness inspire a collaborative approach to leadership which uplifts the presence and voices of marginalized communities. Coming together in community helps us develop the skills of deep listening and compassionate speech, and learn how to reach harmony when there are diverse views. In this way, it becomes possible to build deep-rooted and inclusive communities which can offer spiritual solidarity, generate truly collective insight, and be a place of refuge and renewal for all.

You can cultivate the spirit of community wherever you are. Like a tree, miracles start small and simple. If it's not possible for you to invest your energies in a meditation sangha, local affinity group, activist network, or NGO with the same aspiration, your community-building can begin right where you live and work, with the people you already spend time with. It may be as simple as gathering a few co-workers, neighbors, or like-minded friends to share a cup of tea and cookies and just taking time to break free from the usual dynamic to speak from the heart and really listen to each other's concerns. All kinds of good things will come from it.

Once, when we were in New York on a teaching tour, a few of us monastics were invited to lead a two-hour mindfulness session for young journalists at the New York offices of the Huffington Post. I remember thinking, "How on earth can we help make a difference in just two hours?" Finally, we decided to offer no more than twenty minutes of guided meditation and relaxation followed by a short, ten-minute talk. And then the rest of the time—an hour and a half—we just listened. We created a space for the young journalists to share deeply from the heart, and we followed our breathing while we heard everything they had to say. We asked them: "Why are you here? Why did you first want to be a journalist, anyway? What's your deepest aspiration? What's your deepest fear? What makes your heart sing?" I remember that some cried while they spoke. At the end of the session, one said it felt like the first time they had truly arrived, into the team, into the office, into their life. Another said it was the first time they really got to hear *who* their colleagues were. Sometimes we just have to drop the mask and let ourselves be human beings, doing our messy best, on a human path, together.

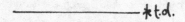

To Engage or to Meditate?

The main task of a community of mindful living is not to organize events—whether they are events for mindfulness practice, or social justice, or engaged action. The main purpose of a sangha is to cultivate brotherhood, sisterhood, and harmony. And, with a sangha like that to take refuge in, everything is possible. We are nourished and we don't lose our hope. That is why mindful communication, deep listening, and loving speech are so important: we need to find ways to keep communication open, share views, and come easily to collective insight and consensus. That is real sangha-building, and it takes time and energy. We need a lot of patience. We need time to sit together, eat together, talk together, and work together, and to cultivate a collective energy of mindfulness, peace, happiness, and compassion.

In this way we can support and nourish each other so we can continue for a long time and not burn out. Your community is your refuge. And, even if it still has its weaknesses, it's always going in the direction of generating more awareness, understanding, and love.

When I was representing the Vietnamese Buddhist Peace Delegation at the Paris Peace Talks, many young people volunteered to help us. We'd work together and share simple meals. In the evening, they would stay on with us to practice sitting meditation or walking meditation or deep relaxation and singing. Soon we began to organize sitting meditations at the nearby Quaker Meeting House. Being in contact with young activists doing peace work and social work, I saw their difficulties. I saw how easy it was to burn out and give up. I knew that, without the practice of sitting meditation, walking meditation, mindful eating, and working together, I could not survive. And so, community-building became a kind of medicine to survive. It's not true that Engaged Buddhism is only when you are engaged in social action. Engaged Buddhism takes place at any time, whether you are walking, sitting, or drinking your tea in mindfulness. That is also Engaged Buddhism because you are doing it not only for yourself. You are doing it to preserve yourself in order to help the world.

We may encounter people who are very active but are not operating from their compassion. They are tired. When compassion is weak, it's not possible to be happy; you become easily jealous, frustrated, and angry. We have to know our limits. You cannot do more than you can do or you will burn out. We have to organize our life to ensure balance. Working with a community, we receive the collective energy of support. Our friends help us not lose ourselves in the work. From time to time we can step back while others step forward. You need the courage to say no, or you will lose yourself very soon. And that is a loss for others and the world. As a teacher, one of the hardest things for me to do is to say no to requests to lead retreats around the world. I know the retreats can benefit many people. But we have to see our limits. Preserving ourselves is a way to preserve our opportunity to serve others.

Engaged Mindfulness in Action

—T.D.

Cheri Maples, the police officer from Madison, Wisconsin, found ways to resource herself and her aspiration with the energy of mindfulness and the spirit of community. After attending her first retreat with Thay in the early 1990s, she joined a local meditation sangha, and stuck with it. Cheri found ways both to resource her own spirits with community connection and to actively nurture community as a field for change. Cheri saw community not only in terms of her local meditation sangha but *wherever* she was, whether it was her workplace or with her family. Cheri focused on three areas: her own inner work (meditation and mindfulness practice, which she called "the foundation for everything"), her relationships, and her engaged practice. She discovered a need to find what she called her own "Zen activities" and invest time and energy in them. These are "activities that completely absorb you, that develop the same things mindfulness does: concentration, focus, finding the extraordinary in the ordinary." For Cheri it was baseball.

As a police officer going out on calls, Cheri started to take more time and to develop a different approach to conflict resolution. There was one call where a divorced dad was refusing to release his young daughter back to his ex-wife. When he answered the door, Cheri, five-foot-three, found herself facing a very angry man who was six-foot-four. He threatened her. "But," she said, "I could just *see* his suffering. It was so obvious." Instead of arresting him on the spot, she asked if she could come in and talk. "I violated every policy in the book, and with no backup, and wearing my gun belt and bulletproof vest, I sat down on the couch next to this guy, which you're never supposed to do. And he just started crying in my arms."

The girl was returned to her mother, and no arrest was needed. Three days later, Cheri ran into the man on the street. Or rather, he ran into her, picked her up in his arms, and gave her a bear hug, saying, "You! You saved my life that night!" Within a few years, Cheri was leading the police department training program, and leading meditation programs for criminal justice professionals, judges, attorneys, prison officials, and social workers.

In her teams, Cheri facilitated decision-making by consensus wherever possible and invested time in thinking about how to integrate the Five Mindfulness Trainings, in a non-sectarian way, into the training of new recruits. She explained, "We have to show them and the rest of the organization that we create the community together. It's not about what the people above them or below them are or aren't doing; it is about what we're *each* doing as individuals to create the community together. Ethics is now integrated into every single thing we teach. It's not a separate course." Recruits' partners and family members were asked to attend some of the sessions to help the officers learn when police skills like "command presence" and "taking control" are appropriate and when they're not. "If they can't do this," said Cheri, "they will not make very good police officers, and they will make terrible partners, spouses, and parents." Cheri said that she wanted them "to think about *what they want to be* as human beings and *how they want to interact with others* on the planet. I want them to understand that, the more open they are, the more likely they will be to perform this job with the open heart required to be effective."

How can we continue this kind of impactful work without feeling overwhelmed by the obstacles or overstretched by the possibilities? Cheri balanced her professional talents with both a profound inner spiritual strength and engaged external efforts to work for systemic change. Here's just one of her insights about self-compassion: "Burnout is a sign that we're violating our own nature in some way. It's usually regarded as a result of giving too much, but I think it results from trying to give what we don't have—in that sense it's the ultimate in giving too little. But when the gift that we give is an integral and valued part of our own journey, when it comes from the organic reality of inner work, it's going to renew itself and be limitless in nature. And that means we have to keep our practice very strong and very alive."

Cheri proposed insightful ways to reform the criminal justice system, including investigating the root causes of racial profiling, re-examining police standards for deadly force, proposing concrete ways to build trust between officers and communities, and developing new programs to address the emotional resilience and trauma of officers (whether they recognize it or not). As part of Cheri's Engaged Buddhism, she also looked deeply into the unconscious agreements and cultures that exist in organizations, including policing, and found that the insight of interbeing can be both empowering and transformative:

> We seem to believe that someone or something else is the problem, and that someone needs to do something better for things to change. We forget that we're a member of this organization! People come out of a meeting and say, "Oh, that was a terrible meeting." And I say, "Were you there? It was a terrible meeting because we all made it a terrible meeting. What could you have done to improve it?" In authentic community membership, we're always holding ourselves accountable for the well-being of the larger community. We become more than just judging critics and consumers, and we start to believe that this world, this organization, this meeting, this gathering, is ours to construct together.

Cheri passed away in 2017 from medical complications following a bicycle accident. In her life, with deep courage, Cheri made peace in herself and brought peace to the world, and in the dimension of action she continues to shine.

Success and Freedom

Can you be ambitious and mindful? Can you be determined to succeed, yet also live a simple life? The question of power is an important one because many of us abuse our power, even if we don't have a lot of it. Parents may misuse their authority over their children and yet they may still feel powerless, unable to do anything to change or help their children. Power is always limited, including political or economic power. Even the President of the United States of America may feel powerless; even multimillionaires may feel powerless.

What does the Buddha say about power and authority? In Buddhism, we speak of three kinds of power everyone can seek. In Chinese they're known as 三德 (*tam đức* in Vietnamese). There's no danger in seeking these three powers because they are the kind of power that can make you and others happy. It's a different kind of power from wealth, fame, influence, and sex.

The first power is *the power of cutting off* (斷德, *đoạn đức*). It is the power to cut off your cravings, anger, fear, despair, or jealousy, which are like flames that ignite and burn you up. As soon as you can cut yourself free from them, you will be very happy. When we crave objects of desire, we're like a fish biting the bait. And the bait has a hook in it. Many of us cannot see the hook in what we're craving, and we get caught. But, with the sword of understanding, you'll be able to see the danger, the hook, in whatever you are craving, and you'll be able to cut off the craving. The same is true with our anger, our jealousy—we use the sword of understanding to cut ourselves free.

The second power is *the power of understanding* (智德, *trí đức*). If you have enough mindfulness, you cultivate concentration. And, with mindfulness and concentration, you can look deeply and get a breakthrough into the heart of reality. You can liberate yourself from wrong views, misunderstanding, and wrong perceptions. And you get free.

Manjushri, the bodhisattva of great understanding, is always depicted holding the sword of wisdom. With that sword he can cut through all kinds of misunderstanding. As a meditator, a practitioner of mindfulness, it is with *wisdom* that you can solve your difficulties. You become rich in insight, rich in freedom. Nobody can steal it from you; nobody can use a gun to rob you of your wisdom.

So, the first spiritual power—the power to cut off—sets you free from craving and anger. The second spiritual power—the power to understand—helps you remove delusion and misunderstanding. And the third spiritual power is love (恩德, *ân đức*). This is *the power to love, forgive, and accept others* and offer understanding and love. There are those of us who cannot accept the other person or we have a hard time accepting the situation as it is. We say, "If they won't change, why should I? If they continue to be like that, I have the right to continue to be like this." But, as soon as we accept them as they are, as soon as we accept the situation, we are free to move on. We stop reacting, and we begin to truly act. If we are always reacting, it doesn't get us anywhere. But, with the power of love and acceptance, you are free to respond with loving kindness and wisdom and you can change the situation. The capacity to accept and forgive is a tremendous source of power.

If you invest your time in cultivating these three kinds of power, there's no danger at all. And the more power you have, the happier you'll be, and the happier those around you will be. With this kind of power you'll never become a victim of your success. With these three powers, there's no danger in having some wealth or some fame: you will use it to help others, help society, and help the planet. It's not the case that a good meditator must always seek poverty. It's okay to have money, but you should know how to use that money to realize your ideal of compassion and understanding.

In spiritual traditions we speak of "voluntary poverty." You want to live simply so you don't have to spend all your time earning money. You want more time to enjoy other things deeply—whether it's the wonders of the planet or your loved ones. Living simply gives you more time to enjoy life. You may be "poor" but you have chosen to be poor. In fact, you are very rich because everything belongs to you— the sunshine, the blue sky, the birdsong, the mountains.

Every moment of your daily life belongs to you. There are those who are rich in terms of money but they don't have the sky, the mountains, or time to take care of their loved ones. Buddhist teachings are very clear on this. Buddhism is not against having money or a good position. If you have true spiritual power—if you have the power to cut off your craving and if you have insight and love—then you have a lot of freedom and happiness. And whatever money, power, or influence you do have may help you realize your ideal of a bodhisattva.

The Art of Power

Don't think that, if you don't have money or position, you're helpless and can't do great things. I have met many wealthy and powerful people who suffer deeply, and many of them are unable to help others. They are so concerned with accumulating wealth that they don't have time for themselves or for their family. I myself am very happy without wealth and power and, in fact, I've been able to help many people. There are things I can do that others cannot. I can go ten days without food. Or, when I hear something insulting, I don't get angry; I'm able to smile. Many people cannot do that.

Please don't think that if you don't have money you can't do anything. It's simply not true. If you are free, you can do so many things to help your people, to help your community. And, when you transform yourself into a bodhisattva, you have a lot of power—the kind of power that helps you be free and helps you bring relief to many people.

There are those who try to get political power at any cost because they believe, without it, they won't have anything. But if you try to get power by destroying your own values, then you lose yourself and you lose the trust of your people. That is why we should not try to get power at any cost. Even without power, you can strengthen the foundations and work at the grassroots, where you can cultivate more trust, love, and solidarity, and influence the situation. And, when your side gets stronger and it's your turn to be in power, you don't need it too much; otherwise, you get corrupted. Real power should always have a spiritual dimension.

Mindfulness Is Not a Tool, It's a Path

A journalist once asked me, "Is it right to bring mindfulness into corporations, helping them have greater success and profit? Is it right to use mindfulness to help the rich get richer? Is that really mindfulness?" There are others who ask whether it's right to teach mindfulness to the military. They say it's one thing to use mindfulness to help veterans but quite another to use it to help soldiers in active service. How can it be ethical to train someone in mindfulness in order to kill better? Is mindfulness being exploited for the wrong aims?

The question is whether mindfulness can benefit everyone or only certain categories of people. Can we *exclude* corporate leaders or the military from the practice of mindfulness? What about other professions such as fishermen? Fishermen also destroy a lot of life as do weapons manufacturers or industrial meat farmers. Should we exclude all of them?

Our Plum Village practice center in France was one of the first to offer mindfulness retreats to businesspeople in the 1990s because we know they suffer just like the rest of us. Two thousand five hundred years ago, the Buddha also taught businesspeople.

Right mindfulness is, first of all, not a tool or an instrument but a path. Right mindfulness is not a means that can be used to arrive at an end. A tool is something that can be used in different ways, like a knife. If you give someone a knife he might use it to chop wood or cut vegetables, but he could also use it to kill or steal. Mindfulness is not like a knife. It's not a tool that can do either good or bad. And yet many of us speak of mindfulness as a tool. We say that with mindfulness we can heal, with mindfulness we can reconcile, with mindfulness we can make more money, with mindfulness we can kill the enemy more effectively.

True mindfulness is not only a path *leading to* happiness but it is a path *of* happiness. When you practice breathing in mindfully, your in-breath is not a means to an end. If you know how to breathe, then you get pleasure, peace, and healing right away while breathing. If you suffer while breathing in—if you have a tendency to think, "I'm suffering now so I can experience something better later on"—that's not right mindfulness. In right mindfulness every step of the path is the path itself. We need to keep reminding ourselves to practice in such a way that we get peace, calm, and joy right away.

Is It Ethical to Teach
Mindfulness to the Military?

In a confrontation during the war in Indochina, the commanders of both the communist and anticommunist armies ordered their soldiers to attack. But the soldiers didn't want to kill each other. The two sides were bunkered down on either side of a river but wouldn't fire. They stayed there quietly for several hours and then, to do their duty, fired their guns into the air, ate their lunch, and went home. This happened time and again in Laos and has happened throughout the history of war. These soldiers had insight. They didn't understand why they had to kill and be killed.

The soldiers could see that the other people were not their enemy; they were just like them, pushed into the front line to kill and be killed. There was mindfulness, and wherever there is mindfulness, there is insight. The insight here is that the other side are also victims of the war. The soldiers who refused to fire on each other had mindfulness, and they also had insight. The soldiers could see the true nature of the situation they were in, and they could see the preciousness of life. It upset their commanders a lot.

Military leaders of our time are trying to train soldiers *not* to have that kind of insight and mindfulness. They think their soldiers can learn mindfulness and concentration as tools, to use only during deployment, in order to be more calm and concentrated and perform the act of killing better. But it's not true. When you instruct someone in right mindfulness they learn how to breathe, how to walk, how to be aware of their feelings and emotions, how to be aware of the fear and the anger inside and around them. When soldiers are aware of these things, they get insight, and that insight will always help them avoid wrong thinking, wrong speech, and wrong action.

If someone goes as a mindfulness teacher to help military leaders realize their aim of doing a better job of destroying the enemy, they're not teaching true mindfulness. Suppose a soldier has been deployed and is at his post. He practices breathing in and out in order to be aware that his enemy is hidden somewhere right there in front of him. He is mindful. "Breathing in, I know the enemy is there. Breathing out, I know I have to kill him before he kills me." The soldier is motivated by fear, by the will to survive, and by a wrong view. He has been trained to see the other person as evil, as the enemy of his nation, as a threat to national security. He believes the world will be better off without him. As a soldier he's been trained to think like that so he'll have the will to kill.

A mindfulness teacher cannot train people to have more mindfulness and concentration in order to kill better. We cannot teach a soldier "Breathing in, I know my enemy is there. Breathing out, I pull the trigger." That's training them in wrong mindfulness, in mindfulness as a tool rather than as a path, in mindfulness without insight. You don't need to practice for ten days or ten years to get insight. With one mindful breath we can get the insight that life is precious. If you instruct a soldier in right mindfulness, he'll have insight, he'll have right view. Once he has right view, he cannot do wrong. Therefore, there's nothing dangerous about teaching soldiers right mindfulness.

The military suffers, and that's why they need help. There is nothing wrong in helping them to suffer less. Right mindfulness is what you can offer. Today we have professional armies, and a young person enlists in the army because they're looking for a salary and a career. They're looking for better prospects than they have as a civilian. They think an army career will make them happy.

If, as a soldier, they can learn true mindfulness, they'll discover what can make them truly happy. They'll be able to recognize and embrace their fear, anger, and despair and they will suffer less. Once they've tasted true happiness they'll realize something about their life and they will change their motivation. It will happen slowly and gradually. As their teacher, you don't urge them to abandon their job. You come only in order to help them to suffer less. And when they suffer less and see what true happiness is, then everything else will change by itself.

When military and political leaders have wrong views, millions of people can be killed, millions of lives destroyed. During the Vietnam War, GIs were instructed: "Communism is dangerous. If Vietnam is allowed to fall under the communists, communism will spread all over Southeast Asia, then to New Zealand and Australia, and very soon to the United States." This was a wrong view motivated by fear. Today, America is doing business with communist Vietnam.

Looking back, we can see that pouring all that money, poison, weapons, and so many human lives into Vietnam was not at all intelligent. Wrong views at the highest levels led to an immense amount of death and destruction. A better view, a much smarter approach, would have been to offer assistance to both South and North Vietnam, to communists and anticommunists, to help rebuild the country, recover the economy, invest in education, and so forth. America would have spent much less money helping both the North and the South become happier countries, and they would have won many friends. Insight—right view—gives rise to right action.

So not only do we need to help soldiers have more right mindfulness and insight but we also need to help their superiors: the commanders, the chiefs of staff, the Pentagon, and the policy makers. We should not exclude anyone from the practice of mindfulness. As long as our political leaders continue to have wrong views about national security and national interests, then many young people will continue to be victims of war and be forced to kill and be killed. There are far better ways to secure our national interests and security than using violent means.

I think that if I were invited to guide and teach the army, I would go because I know that the true practice of mindfulness will change the world. It will change soldiers' ideas about happiness; it will change their way of life.

True Mindfulness Contains the Seed of Ethics

In 2013, one hundred Plum Village monastics went to lead a day of mindfulness practice for over seven hundred Google employees at their California headquarters. We know that the young Googlers are hardworking and intelligent. They're under pressure to get new insights and to innovate in order to be the best and to help their company be number one. Their intention is to have more and more success. And our intention was to give them a taste of true happiness. We can't say, "Dear friends, you have to abandon your motivation first, before we'll teach you mindfulness." They want to be successful, and perhaps they want to learn mindfulness because they think it will help them be more successful.

As long as we teach right mindfulness, we needn't be afraid. With right mindfulness, anyone can taste true happiness, freedom, and love, and their intention will naturally change. Instead of wanting to become number one, they will want to be truly happy. A healthy intention to experience true happiness and offer it to others, an intention to live and work in such a way that helps people suffer less, can bring a lot of joy and make the world a more beautiful place.

When the Buddha taught right mindfulness, he always taught it as one element of the noble "eightfold path," the path of happiness and well-being expressed in the Five Mindfulness Trainings. You can never take right mindfulness out of the context of the eightfold path. If you do, it's no longer true mindfulness. Right mindfulness cannot be separated from right concentration, right view (or "insight"), right thinking, right speech, right action, right livelihood, and right diligence. Right mindfulness has the nature of interbeing with all these other elements of the path. If you have not seen the other seven elements of mindfulness, you have not really seen mindfulness.

Many of us have a tendency to think dualistically, and we apply that thinking to spiritual traditions whose teachings *transcend* dualistic thinking. We need to be aware that this may happen to mindfulness. We need to train and practice mindfulness *as a path of happiness and transformation that is deeply connected to all the other elements of the path.* Mindfulness cannot be separated from the mindfulness trainings and the practice of applied ethics.

We need to train thousands of mindfulness teachers, hundreds of thousands of teachers, because everywhere needs the practice of mindfulness. We don't need to be a Buddhist to practice right mindfulness. These teachings and practices are the heritage of the whole of humankind, not only Buddhists.

And we should always remember that what we call "Buddhism" is made *only* of non-Buddhist elements.

Whether mindfulness stays right mindfulness or not depends on our own training and practice. The practice, provided it is *true* practice, can only help. Don't be afraid. Each one of us has the capacity to master this practice and bring joy, healing, and reconciliation to the world.

Communities of Resilience

—T.D.

How can the insights of right mindfulness help us heal and transform the violence, inequality, and systemic injustice of our times? There is, as Dr. Larry Ward says, a profound "human crisis" at the center of our Earth's crisis; there is a racial karma that is calling for healing. The fact is, we continue to harm and discriminate against the planet because we continue to harm and discriminate against each other. The two are deeply interconnected. As Larry explains in his book, *America's Racial Karma*, our task is to walk a path of mercy and healing, "so our racialized consciousness may become profoundly humanized, to care for ourselves and our planet."

There is inner and outer work for us all to do. We need to work to transform policies and systems of institutionalized race consciousness in our society, and we also need to do the inner work:

to have the courage to develop our own embodied, applied, spiritual practice.

As Larry explains, "Our bodies hold the retribution energies of America's racial karma. No one escapes this fear and trembling deep in our bones. Whether we were—or are—victims, perpetrators, or witnesses, we are unavoidably biologically destabilized or dysregulated by our sensory experience or the memories of it." The work of our spiritual practice is to recognize, embrace, and heal this trauma in our own bodies, heart, and mind, and to help others do the same. Without this transformation at the roots, Larry says, the potential for deep, systemic change will be blocked.

Combining inner practice and outer transformation, Larry invites us to create what he calls "communities of resilience"— communities where we make a conscious intent to live together "in kindness, openness, generosity, sanity, and love." These actions, he says, must manifest in concrete, embodied ways. Healing is an art, and we need to help one another on the journey; we need to create moments and environments for the healing of our own pains and traumas and for the healing of our family, friends, co-workers, and fellow citizens.

When he visited the US Congress, Thay proposed establishing a "Council of Sages" to conduct sessions of deep listening for the nation. Those who are wise and loving, spiritual leaders with the capacity of listening with great compassion, can be invited to join the council, and they can create a safe environment for those in society who feel they have been victims of discrimination and injustice to speak out and express themselves. The sessions could be broadcast live. It may take days or weeks for those who have suffered to get enough courage to share everything that is in their heart. "Some call me idealistic," Thay said. "And yet practicing compassion is something we *have* to do, in order to get out of situations of anger. It's the only way out. It is a universal door."

As Larry says, "The bridge of mercy lies deep within us and among us, however well it is hidden by clouds of conflict, cruelty, and hatred . . . Karma can be healed, and karma can be transformed, but only if we choose to turn the wheel to another track." Larry offers these three powerful mantras, flowers from his heart's garden, to support a deep, embodied, enlivened path of racial healing:

Stand up in the house of belonging

Don't act like this is not your land. Don't act like you can't take charge because it's obvious to me that the principalities and powers who are supposed to be in charge of this land at this moment are absolutely incapable. So, stand up! Act like you are a real human being. Don't let the messengers of systemic racism define your life for you. Don't let them define your power for you.

Take your seat at the table of healing and transformation

My grandma said, "Don't let some fool take your seat." Take your seat. Be present and care for yourself; love yourself. As you love yourself and care for yourself, that love will move outward. It will spill out all around you with a fragrance of holiness.

Ride the winds of change, unafraid

Act like the mighty ones of old who knew no fear. Embrace their wild resilience and their vision of what is really possible for us together in healing America's racial karma.

These vivid mantras of freedom and empowerment call on each of us to contribute our part in giving space and voice to those whose pleas for justice and equality are being silenced or ignored. In the light of interbeing, each one of us has a role to play in transforming our racialized consciousness. There is, as Larry says, both inner and outer work for us all to do.

 *t.d.

ZEN AND THE ART OF SAVING THE PLANET

The World as a Koan

With a community, we support each other to look deeply into our real situation to try to find a way out. In the Zen circle there is a practice of koans. A koan should be something that you are deeply interested in—your deepest concern. You want to understand. You want to transform. Holding a koan in your heart and mind is like being struck by an arrow. Standing or sitting, awake or sleeping, you carry the arrow in your flesh. A koan should be like that. You hold it day and night, embracing it, looking deeply into it. And one day insight will come, and you'll understand, and you'll be liberated.

It is possible to take the suffering in the Middle East, or racial injustice, or the suffering of the planet as a koan for the whole of humankind. Yet as a human family we have been too busy to do so. As individuals and as a society, we pay some attention to these urgent issues, and then we get distracted by the next crisis, the next big problem.

You do not just use your intellect in order to work on a koan. A koan should be buried deep in the soil of your mind. You should be able to mobilize all your strength, all your energy, all your mindfulness and concentration in order to embrace the difficulty, the situation, the deep suffering that is your koan. Day and night, at every moment, you do only that: embracing deeply, tenderly.

One day you will get a breakthrough. The insight may come from you, or it may be an expression of the collective insight.

When the practice of a koan takes place at the community level, it's very powerful. That is why, when we organize a conference in the Buddhist spirit, it has to be organized in the form of a retreat, with time for sitting meditation and silent walking meditation. There must be hours of deep contemplation to replenish our insight. We open ourselves to reality, we embrace what we hear and experience, and with that energy of concentration, there will be more insight.

Individual or Collective?

As I see it, the whole of the twentieth-century was characterized by individualism: everyone for themselves. How I have been training my students is quite different. We don't train as individuals; we train to develop communities. We learn to live together, do things together, and cultivate awakening together. Whatever we do, we do together. If the next generation can be different from previous generations, it will be because you know how to do things together. My deepest desire is that the young generation *can* be different—that you can learn how to be together and act together. Whatever you do can be done in the spirit of community.

Just as a lotus flower is made of many molecules, and all the molecules come together in harmony to produce beautiful leaves and flowers, so, too, can a community bring together all the particular individuals to bring about a universal whole.

If you want to produce a society that is peaceful, happy, and compassionate, you have to visualize the universal community. You learn to go as a river. If you can do that, you will change the world.

There's a tendency to resist the idea of a community as an organism, because we still want to hold on to our person, our self. We're not yet ready to live the life of a cell in the body of the community. This takes quite a turn, quite a transformation. In my own life, the more I reflected and looked deeply into the Buddha's wisdom, and the way he organized his community, the more clearly I saw the path of practice.

The moment I got that insight, I received new eyes. I looked at my friends and students in a very different way. I saw that *I am them, and they are me*. And I saw that everything I do, think, and say is for nourishing and transmitting insight to them. In the future, whether I am there or not, it is no longer a problem because I have penetrated the insight of no-self. There is no longer any discrimination between myself and others, no longer any resistance. You accept others as you accept yourself. And, in that kind of relationship, you can have a lot of happiness.

We want to have a young community that is able to transform the world and to protect Mother Earth; able to reduce suffering and promote more physical and mental health; able to bring the practice into schools, corporations, and even the army. It is possible for us to bring mindfulness everywhere, not as a religion but as a practice that can bring relief to everyone in society.

Wake Up for a Future to Be Possible

If we want to help society transform intolerance, discrimination, craving, anger, and despair, we can do so with the ethical guidelines of the Five Mindfulness Trainings: a concrete practice of true love and compassion, clearly showing the way toward a life in harmony with each other and with the Earth.

Many of us feel anger and frustration when we see environmental destruction, injustice, and inequality, and we feel despair because we don't seem to be strong enough individually to change our way of life. Coming together in community offers a way to pool our energy and act in synchrony. Our collective practice can bring transformation and healing to ourselves and to society.

Buddhism is a source of wisdom, a long tradition of the practice of understanding and love and not just of devotion. The spirit of the Dharma is very close to the spirit of science; both help us cultivate an open and non-discriminating mind. Anyone can contribute to collective awakening, whatever your culture, spiritual roots, or beliefs. The practice of maitrī, of loving kindness, friendship, and togetherness, is at the foundation of the path.

It is only with collective awakening that we will have enough strength to make the changes we need to protect ourselves and the planet. Nothing can be achieved without the energy of togetherness, of brotherhood and sisterhood. It is critical for changing the present moment and changing the future. Brotherhood and sisterhood is a kind of monument, and it takes time to build it. But, with brotherhood and sisterhood, we have hope.

The future belongs to the young generation. You need to wake up. It is possible for you to be something and to do something now, to help. Do not despair. There is *always* something you can do. There is still a chance. So, recognize what you need to do, do it, and you will have peace.

be alive
be the
miracle

EPILOGUE

We Have More Than Enough Bodhisattvas

The Lotus Sutra tells a story about a bodhisattva whose name is Gadgadasvara, which means "Wonderful Sound." He is a musician and composer and serves the world with his music. Legend has it that Gadgadasvara is a bodhisattva from another planet. From time to time, while sitting with his sangha, the Buddha Shakyamuni would use his beam of mindfulness to reach out into the cosmos and get in touch with different worlds. In this way, bodhisattvas and buddhas on other planets came to be aware that on this tiny planet Earth there was a Buddha teaching.

When Bodhisattva Wonderful Sound got a beam of light from Shakyamuni, he looked and saw planet Earth, with the Buddha and his assembly on Vulture Peak, and he wanted to visit. Many other bodhisattvas joined him. In advance of their arrival they produced thousands of huge, beautiful lotus buds all around Vulture Peak. Everyone wondered why these beautiful lotuses suddenly appeared. Shakyamuni Buddha explained, "Well, we have some visitors." And he told them about Bodhisattva Wonderful Sound.

In Plum Village we also practice in music. Music can create harmony within ourselves and the community. Sometimes there are many voices inside all wanting to express themselves, and to speak at once. By concentrating on the music of our breathing, we can appease and harmonize all the voices. When the community practices meditation together, the silence and deep, mindful breathing is a kind of music we enjoy together. We're not doing anything; we just produce our being, our full presence, and become aware of the presence of one another. That is already enough to nourish and heal. And so, music sometimes can be very silent. It can calm things down. It can heal. And Bodhisattva Wonderful Sound is someone who practiced deeply that kind of sacred music over many, many lifetimes.

Bodhisattva Gadgadasvara and his friends manifested themselves at Vulture Peak, paid their respects to the Buddha, and offered the greetings of the Buddha of their land. They noticed that our planet is small and yet still has a lot of suffering. They could see Shakyamuni Buddha working hard to help relieve the suffering, and many of them volunteered to stay and help. They were very kind. But Shakyamuni said, "Thank you for your goodwill, but we do have enough bodhisattvas here to take care of ourselves." And at that moment he looked deeply into the Earth, and suddenly many wondrous bodhisattvas—

hundreds of thousands of them—sprang up from the Earth. Everyone was amazed.

Shakyamuni was helping everyone see the ultimate dimension. In the ultimate dimension, the life span of a buddha is limitless, and your life span is also limitless.

This image in the Lotus Sutra is an artful device to help us touch the ultimate dimension, see ourselves in the ultimate dimension, and see the Buddha, our teacher, in the ultimate dimension, and to realize that we do have enough children of the Earth to take care of the Earth.

Don't be too sure that on Earth there are only children of the Earth. There may be living beings from other planets. Meteorites may have brought nascent life with them. And so, on Earth there may be life that is not exactly born from the Earth but arrived with meteorites and naturalized.

The Lotus Sutra gives a very clear impression that we are a small planet, and that the Buddha is one of the planet's children. He wants to take care of the planet and has many disciples ready to hold the Earth tenderly in their arms and take care of the planet. There is no need to fear that we don't have enough people to take care of our home, planet Earth. We know how to do it. Bodhisattva Avalokiteshvara is a child of the Earth and proves that mankind is quite capable of embracing the Earth's suffering and preserving the beauty and wonders of the planet. We are all children of the Earth and we should take care of each other. We should take care of our environment. And with community and togetherness, this is possible.

This morning the birds joyfully welcome the rising sun.

Do you know, my child, that the white clouds

Are still floating in the vault of the sky?

Where are you now?

In the country of the present moment

The ancient mountain is still there

Although the white-crested wave

Is still reaching for distant shores.

Look again, you will see me in you and in every leaf and flower bud.

If you call my name, you will see me right away.

Where are you going?

The old frangipani tree offers its fragrant flowers this morning.

You and I have never really been apart.

Spring has come.

The pines have put out new shining green needles

And on the edge of the forest

The wild plum trees have burst into flower.

FROM "AT THE EDGE OF THE FOREST"

By Thich Nhat Hanh

AFTERWORD
YOU ARE THE FUTURE

Sister Chan Khong

I have been young like many of you, and full of determination to change the situation of suffering in myself, my family, and the world. And yet, often when I achieved what I thought was "right," it was at great cost to myself and my close relationships. It was only when I met Thay that I learned how to handle those difficult moments when I felt lost and overwhelmed by anger, fear, and despair. Thay taught me to always remember to come back to my breathing, and dwell only with the breathing, from within. In this way we can be our best: we can *be stillness* and give rise to a clear mind. Then, deep awakening and compassion can manifest, right in that moment, in our own hearts, and it becomes possible to see it and touch it even in the heart of our so-called "enemy."

In that very moment when you are lost in anger, fear, or despair, please remember that awakening and compassion are always there in you, right in that moment. It is possible to get in touch with what is sacred in you—whether you call it God, Allah, Brahma, or buddha-nature. You get in touch with this energy by coming back right away to your mindful in-breath and out-breath, and remaining silent, without doing anything, or saying anything—even without thinking. Be only with your in-breath and out-breath for a few moments, and you will be able to touch that reality of peace, compassion, and clarity of mind that is already there deep inside you.

This seed of awakening and love is there in you and in all people and all species on Earth. Sometimes we forget. That seed may be lost deep in our consciousness. But the longer you can stay peacefully with your in-breath and out-breath, a safe haven of peace and compassion will grow in you. As you breathe in, you touch deeply that seed of compassion and loving kindness in you; and as you breathe out, you radiate that energy of compassion and loving kindness to those around you, and to the world. This is the energy of Bodhisattva Avalokiteshvara manifesting in you.

I will never forget that early morning in Vietnam, when I found the four bodies of my friends shot on the banks of the Saigon River. In that moment, I was overwhelmed by anger, fear, and despair. But I was able to bring my attention to my breathing for several hours, without trying to think, blame, scream, or curse. I invoked the name of Bodhisattva Avalokiteshvara, and did my best to touch the seeds of love, peace, and compassion in me. It was not easy. The despair was so overwhelming. But I kept returning to my breathing and slowly a deep stillness and calm grew in me. A gentle peace flooded my heart, and I found a way with my colleagues to respond to the attackers with love, understanding, and forgiveness. With stillness in me, I realized that they were only acting under orders; they did not want to do it, but they were forced to. When we spoke at the funeral of our beloved friends, the attackers' informers were there, and the love in our hearts touched the love in theirs, and they never attacked us again. Since that day in all our engaged work, we have always met many bodhisattvas along our path.

If I can do it, you can do it too. Whenever you receive devastating news, or witness injustice, or feel helpless and full of despair, please remember first of all to come back to your mindful in-breath and out-breath. Don't do anything or say anything until you have touched that calm, that peace, that love.

Mother Earth needs you right now. She is calling out for your help. You are her beloved children, and she needs you to be love, to be light, to be peace. You have light in you. You have the energy of the bodhisattvas in you. With a spiritual dimension in your life, you will be able to keep balance and live deeply in every moment, cherishing this life you have to live. And with that energy, you can take action to protect the planet and protect each other. Together, you can do it. Do not be a lone warrior. Find your allies and build community wherever you are. Mother Earth and our spiritual ancestors and land ancestors are counting on you. They are transmitting to you their energy of love and trust, and will accompany you every step of the way.

ACKNOWLEDGMENTS

This book is the fruit of a diverse and vibrant spiritual community working together to bring Thay's rich body of teachings into print. Since his stroke in 2014, Thay has been a warrior, a silent sage, and a boundless source of love, trust, and support as we continue his work. The deepest bow of appreciation and respect we offer first to Thay, and to all our ancestral teachers, for showing our generation the way forward.

We would like to thank the core team of editors in Plum Village who worked together with Sister True Dedication to contribute their skillful edits, deep insights, creative guidance, and bold vision to the book, helping select Thay's teachings and develop the commentary: Brother Phap Dung ("Architect Monk"), Sister Lang Nghiem ("Sister Hero"), Brother Phap Linh ("Brother Spirit"), and Jo Confino. If the arrow of this book has come close to its mark, it is thanks to them.

It is no small task to research and edit a manuscript while actively engaged in monastic life, and we are deeply grateful to the wider community for their support and trust. In particular, it is thanks to our Vietnamese monastics, who have helped bring a liv-

ACKNOWLEDGMENTS

ing tradition of Vietnamese Zen to the West, that we can make these authentic teachings accessible to a new generation today. We would also like to thank the many bodhisattvas of the Wake Up community, the Earth Holder Sangha, and the A.R.I.S.E. Sangha for pioneering new ways to realize climate justice, spiritual practice, community-building, and planetary healing. Their inspiring example has informed the whole book, and it was in guiding them that Thay gave many of the powerful teachings contained in these pages.

For their deep teachings which enriched the commentary, we would like to thank Sister Chan Khong, Sister Chan Duc, Sister Jina, Sister Kinh Nghiem, Sister The Nghiem, Brother Phap Huu, Brother Phap Lai, Brother Phap Luu, Dr. Larry Ward, Cheri Maples, Jerker Fredriksson, John Bell, Glen Schneider, Kaira Jewel Lingo, and Christiana Figueres. For generously offering love, trust, and encouragement in the final stages of completing the manuscript, Sister True Dedication would like to thank Judith and Patrick Phillips, Rebekah Phillips, Sister Huong Nghiem, Sister Thoai Nghiem, Sister Le Nghiem, Sister Luc Nghiem, Sister Tri Nghiem, Sashareen Morgan, Shantum and Gitu Seth, Denise Nguyen, and Paz Perlman.

We would like to express our gratitude to the team at Harper-One and to Gideon Weil for his compassion and positivity, and for having faith in this book from its inception; and to Sam Tatum, Lisa Zuniga, and Yvonne Chan for bringing patience and skill to the book's production and design. Thanks to our community's literary agent, Cecile Barendsma, for guidance and wise advice beyond the call of duty; and to Sister Trai Nghiem, our publishing coordinator, for steering the book to publication with skill, kindness, and grace. We would like to thank our friends at the Thich Nhat Hanh Foundation and Parallax Press for generously sharing resources and for allowing us to quote from Dr. Larry Ward's book, *America's Racial Karma: An Invitation to Heal*, and from interviews with Cheri Maples in *The Mindfulness Bell*. Thanks to Helen Civil and the team at The Resilience Shift for their kind permission to quote (on p. 213–214 and p. 216) from Seth Schultz and Peter Willis's interview with Christiana Figueres in the *Resilient Leadership* podcast episode 23: "What Does the Future of Resilient Leadership Look Like?" And thanks to the team at Climate One from The Commonwealth Club for their kind permission to quote (on p. 215) from Greg Dalton's podcast interview with Christiana Figueres: "A Conversation on Mindfulness and Climate."

Finally, we would like to thank all of you who have come on retreats, been in the audience of Thay's talks, asked him questions, or read his books, and who are actively bringing these teachings into your lives. Because you are there, everything is possible.

ABOUT THE AUTHORS

THICH NHAT HANH (1926–2022) was a world-renowned Buddhist Zen master, poet, author, scholar, and activist for social change, who was nominated for the Nobel Peace Prize by Dr. Martin Luther King Jr. He remains a preeminent figure in contemporary Buddhism, offering teachings that are both deeply rooted in ancient wisdom and accessible to all.

SISTER CHAN KHONG is Thich Nhat Hanh's most senior monastic disciple and lifelong collaborator. A leading force in his engaged Buddhism programs and humanitarian projects, her books include *Learning True Love* and *Beginning Anew*.

SISTER TRUE DEDICATION is a former journalist and monastic Dharma Teacher ordained by Thich Nhat Hanh.

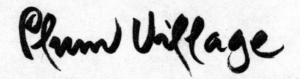

Connect with Thich Nhat Hanh's international community

| For news, online retreats and live mindfulness sessions, visit: **plumvillage.org** | Download the Plum Village App for free meditations and relaxations: **plumvillage.app** |

THICH NHAT HANH FOUNDATION

planting seeds of Compassion

The Thich Nhat Hanh Foundation is the charitable foundation dedicated to continuing the teachings and legacy of Zen Master Thich Nhat Hanh. By becoming a supporter, you join many others who want to learn and share his life-changing practices of mindfulness and engaged Buddhism, and bring change to ourselves, our society and our planet. To find out how you can help support his legacy, and to subscribe to our community newsletter, visit: **tnhf.org**.

Immerse yourself in mindfulness on a residential retreat at one of Thich Nhat Hanh's mindfulness practice centers in the US:

Deer Park Monastery, Escondido, CA: **deerparkmonastery.org**
Magnolia Grove Monastery, Batesville, MS: **magnoliagrovemonastery.org**
Blue Cliff Monastery, Pine Bush, NY: **bluecliffmonastery.org**
To find out more about centers in Europe and Asia: **plumvillage.org**

Discover international networks of Engaged Buddhism in Thich Nhat Hanh's tradition:

Earth Holders
A Mindful Earth Justice Initiative
earthholder.training

ARISE
Awakening through Race, Intersectionality,
and Social Equity
arisesangha.org

The Wake Up Movement
Young Buddhists and non-Buddhists
for a Healthy and Compassionate Society
wkup.org

Wake Up Schools
Cultivating Mindfulness in Education
wakeupschools.org